BrainSnack®
Enjoy thinking!

imagine!
Publishing

Frank Coussement & Peter De Schepper

Brain Busting, Mind Twisting, IQ Crushing Puzzles

With BrainSnack® puzzles

Library of Congress Cataloging-in-Publication Data Is Available

10 9 8 7 6 5 4 3 2 1

An Imagine Book
Published by Charlesbridge
85 Main Street
Watertown, MA 02472
617-926-0329
www.charlesbridge.com

Printed in China, October 2011

Brain Busting, Mind Twisting, IQ Crushing Puzzles
ISBN: 978-1-936140-61-9

For information about custom editions, special sales, premium and corporate purchases, please contact Charlesbridge Publishing, Inc. at specialsales@charlesbridge.com

Binairo®

PeterFrank t.v.,
Postbus 11
B-9830 Sint-Martens-Latem, Belgium
www.peterfrank.be

Contents

Solution strategies

To solve a BrainSnack® puzzle you will have to apply one or more of the following solution strategies. The most important solution strategy is mentioned for many BrainSnack® puzzles.

Look for the elements in the BrainSnack® puzzle that differ from the rest.

Look for similarities and associations between the different elements.

Colors are important.

Pay attention to the three-dimensional design.

Calculate with the digits and numbers that you see and/or discover.

Elements are shifted or follow a logical direction.

Look for the logical order or a repetitive series.

Try to find the logic.

Degree of difficulty

From easy  to difficult 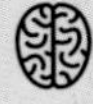   

Introduction

Homo ludens, playful man

One of the questions that pops up regularly is: do brainteasers have a positive impact on brain activity? In other words, are brainteasers a welcome aid in the learning process?

For the last fifteen years, it has been generally accepted that the brain is plastic. In other words, the brain's structure develops constantly as a reaction to the experiences that you gain. The more your brain is stimulated via sports and games, the better your mental condition.

In a nutshell, mental exercises stimulate the brain.

Just like with any sport, you must train a lot and train hard for mental exercises—you should not give up. Some people excel at 3-D puzzles and have trouble solving other puzzles. As with other sports, be sportsmanlike—do not consult the answers at the back. Simply try again later or continue with a different puzzle.

Every puzzle in this book is a brain-busting, mind-twisting, IQ-crushing contest for playful people.

BrainSnack® Puzzles

What remains in a corner and goes around the world?

Solution on page 288

Logical snack

Comparing your brain with a computer is easy, but unfortunate. After all, a computer is designed to execute all procedures in accordance with a strict plan and a logical order.

In comparison, your brain connects a large quantity of parallel information channels, while a computer usually processes information sequentially. The confusion in your brain is related to the evolution that our brains underwent. Each time new brain structures were developed, the existing structures had to connect to the new reality.

Actually, the inside of your head is quite like a busy fast-food restaurant where the cooks seemingly hamper one another, but in the end everyone gets their snack on time.

To like these snacks—the BrainSnack® puzzles—you must think logically. You can use one or several logical strategies like direction, differences and/or similarities, associations, calculations, order, spatial insight, colors, quantities, and distances. A BrainSnack® puzzle ensures that all the brain's capacities are utilized.

You will want more!

BrainSnack® Puzzles

1

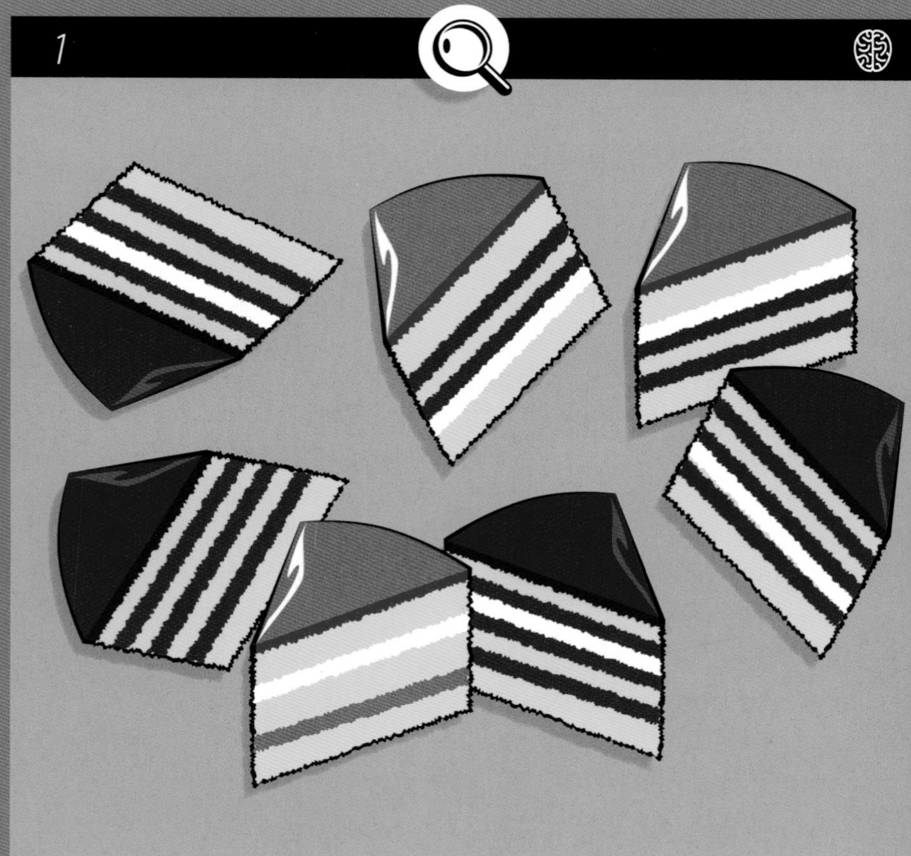

From how many different cakes do these pieces originate?

BrainSnack® Puzzles

2

How many unripe berries are missing from the last branch?

BrainSnack® Puzzles

3

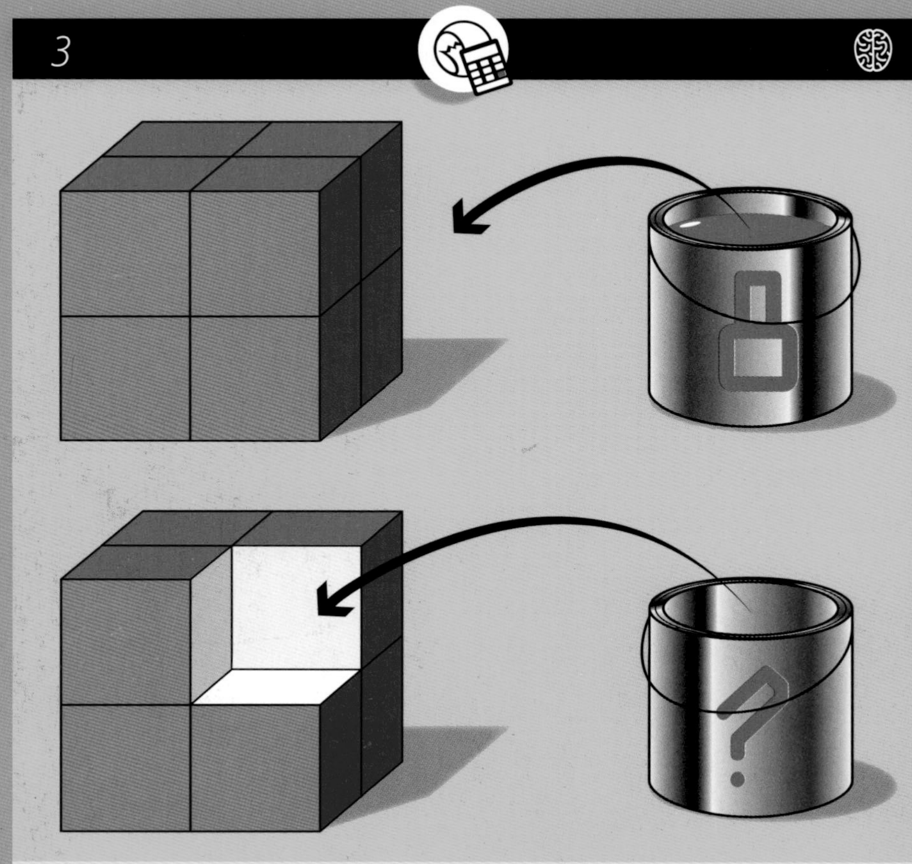

To paint the outside of this cube, you need eight gallons of paint.

How many gallons of paint do you need to paint the white part?

BrainSnack® Puzzles

4  

NEWPORT
EVERSOR
VOVINO
ORLEMAN
RIMIN?

What letter is missing from the name of the last port the yachtsman will enter?

BrainSnack® Puzzles

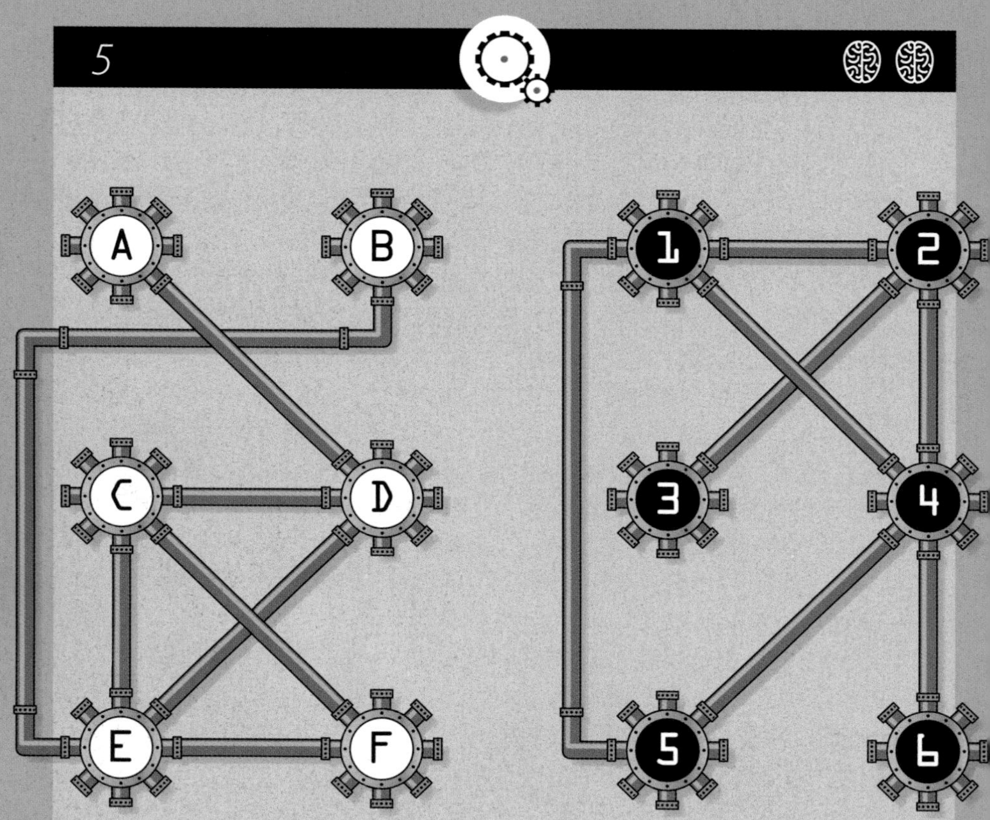

Replace every number on the right picture with a letter (A–F) so that the collectors of the pipes are linked in the same way as on the left picture (from A to D, B to E, etc.).

Answer like this: 1D2A3C4B5F6E.

BrainSnack® Puzzles

6

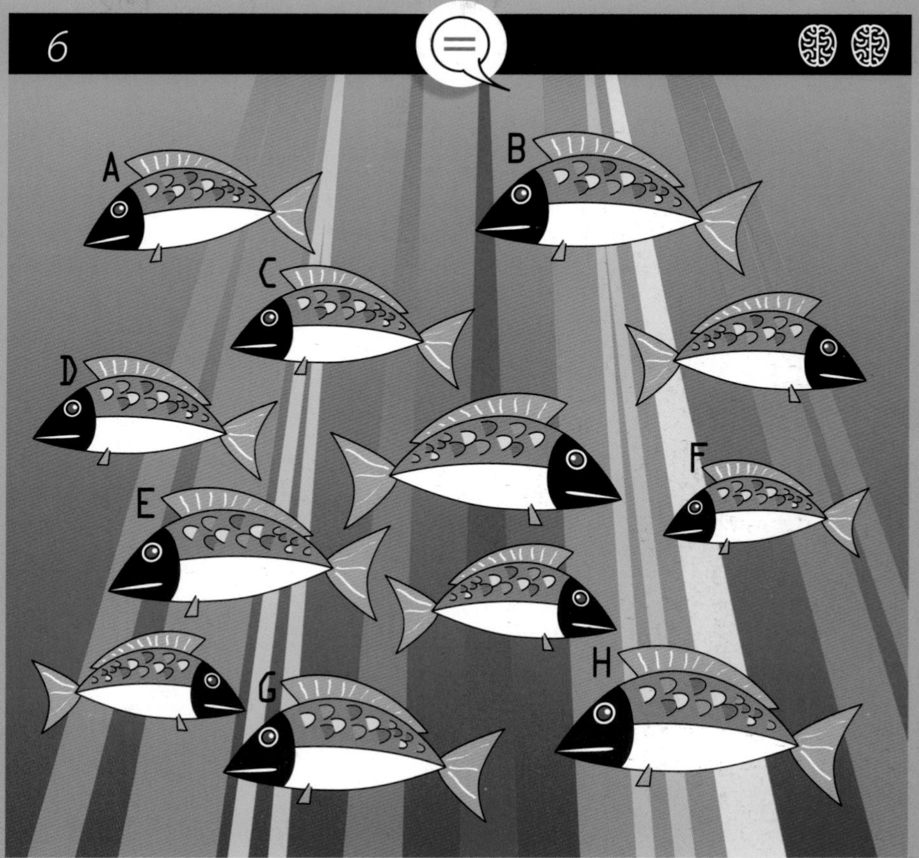

Which fish (A–H) is swimming in the wrong direction?

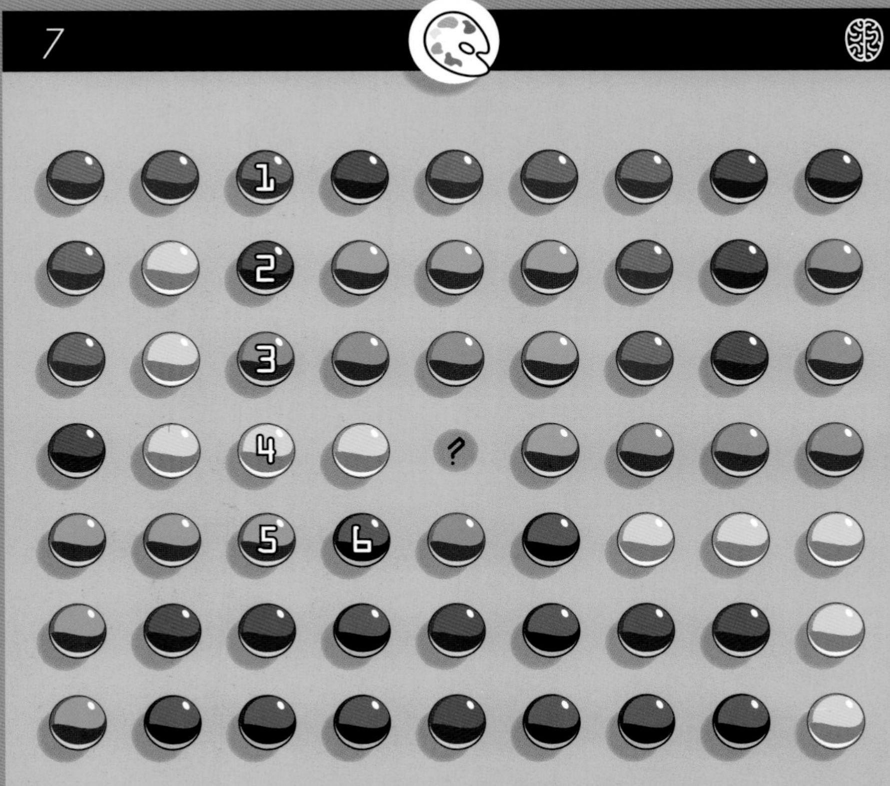

Which marble (1–6) should replace the question mark?

BrainSnack® Puzzles

8

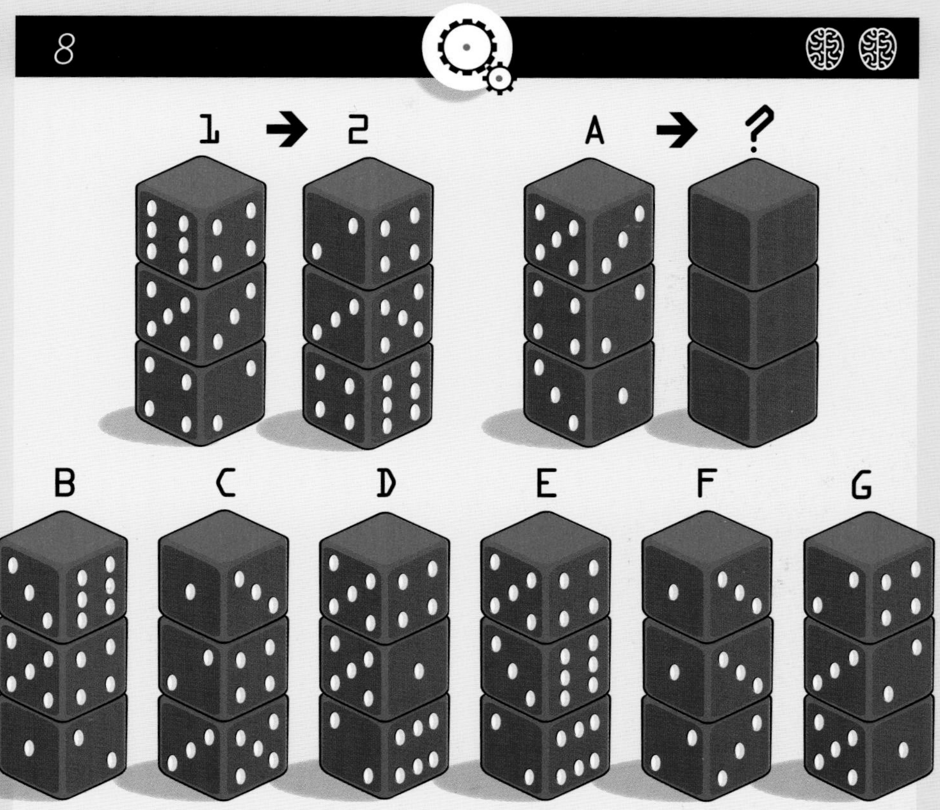

Stack 1 is to stack 2 as stack A is to which stack (B–G)?

BrainSnack® Puzzles

9

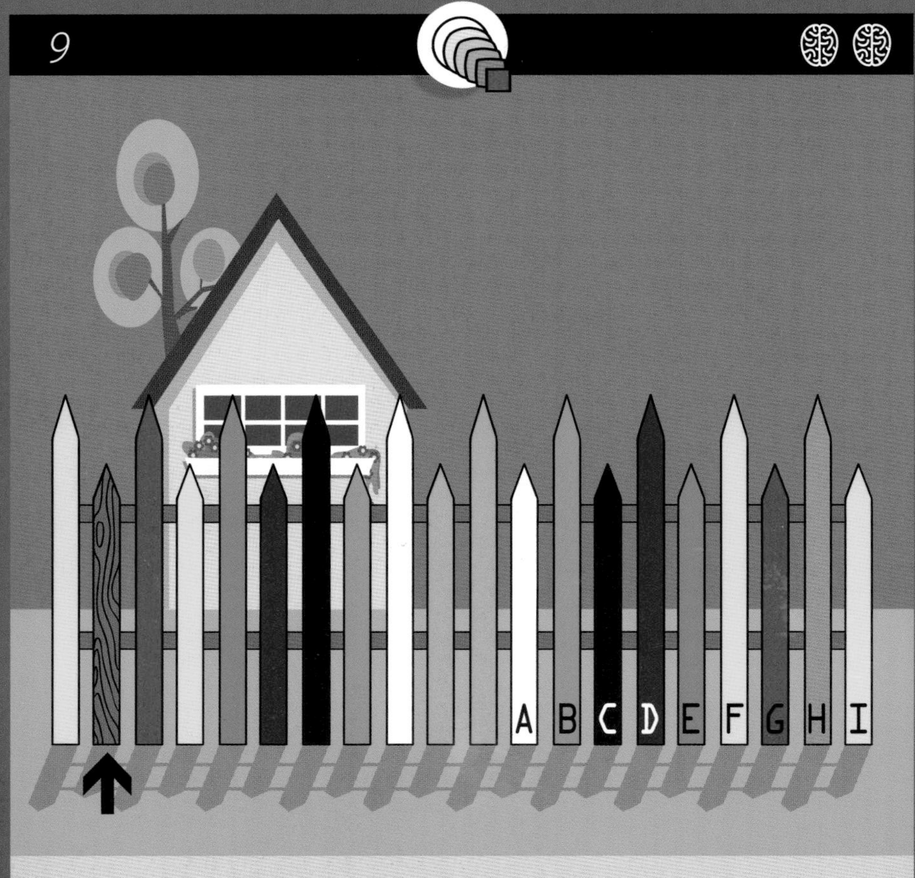

What color (A–I) should the wooden slat be painted?

BrainSnack® Puzzles

10

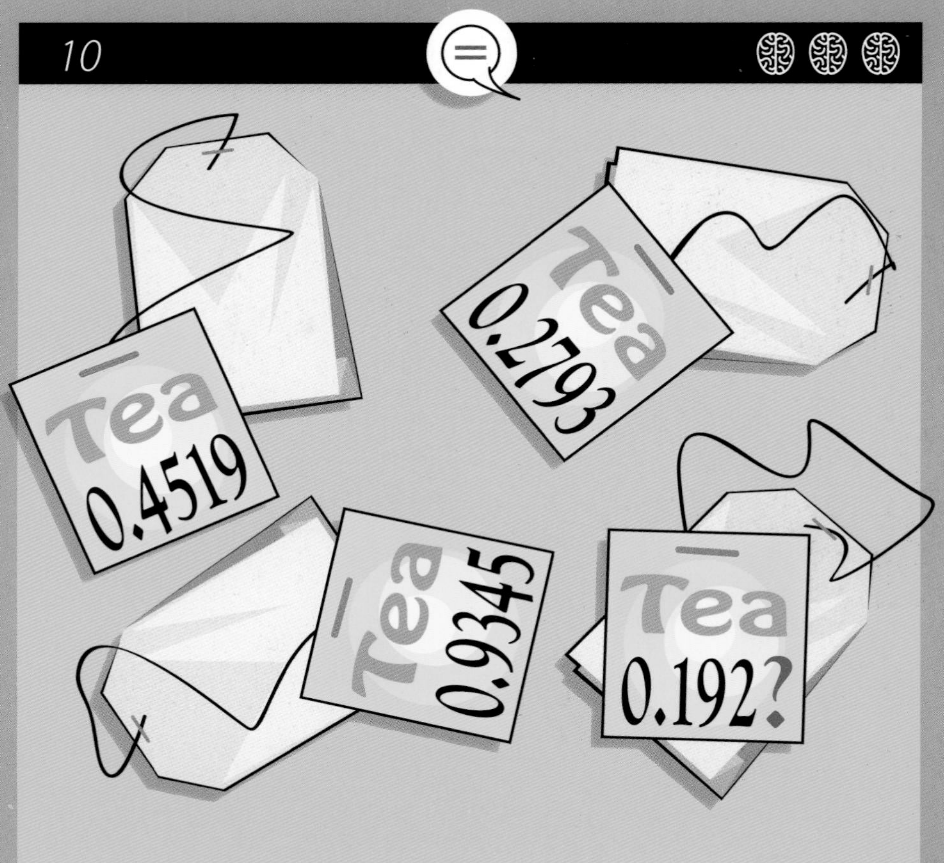

What digit should replace the question mark on the last tea bag?

BrainSnack® Puzzles

11

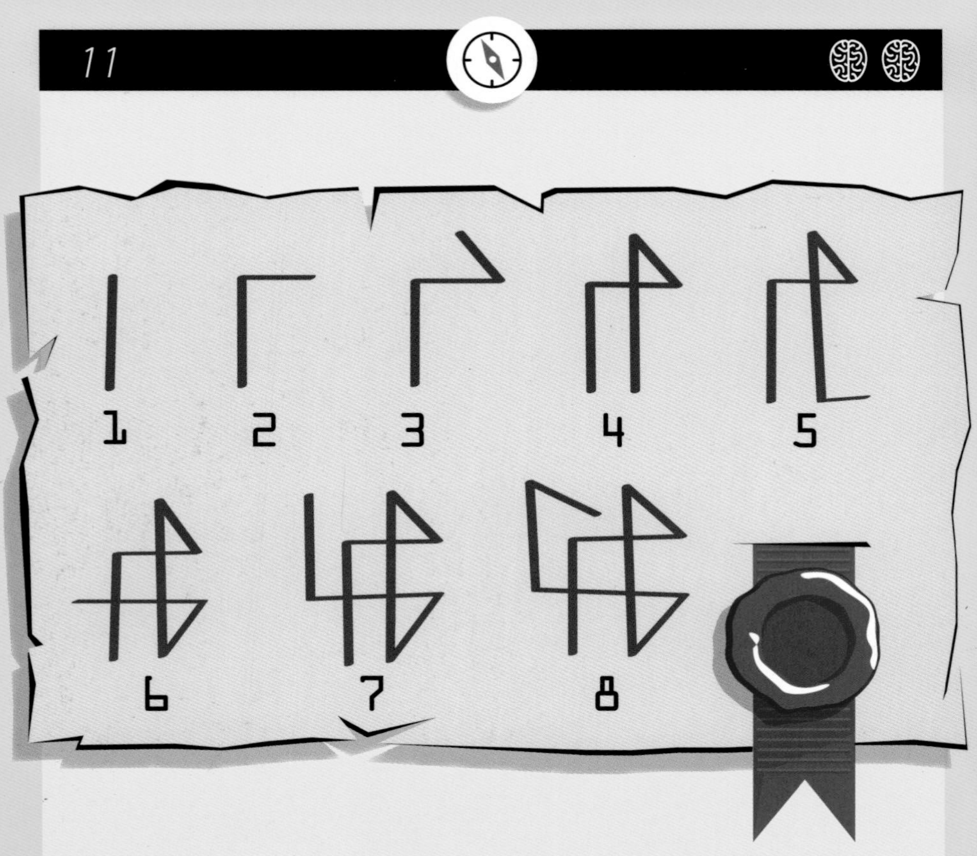

Which symbol (1–8) is drawn incorrectly?

BrainSnack® Puzzles

12

The number on each pair of hearts shows how many years these couples have been in love.

Knowing that there is no logical relationship between the numbers, how long has the last couple been in love?

BrainSnack® Puzzles

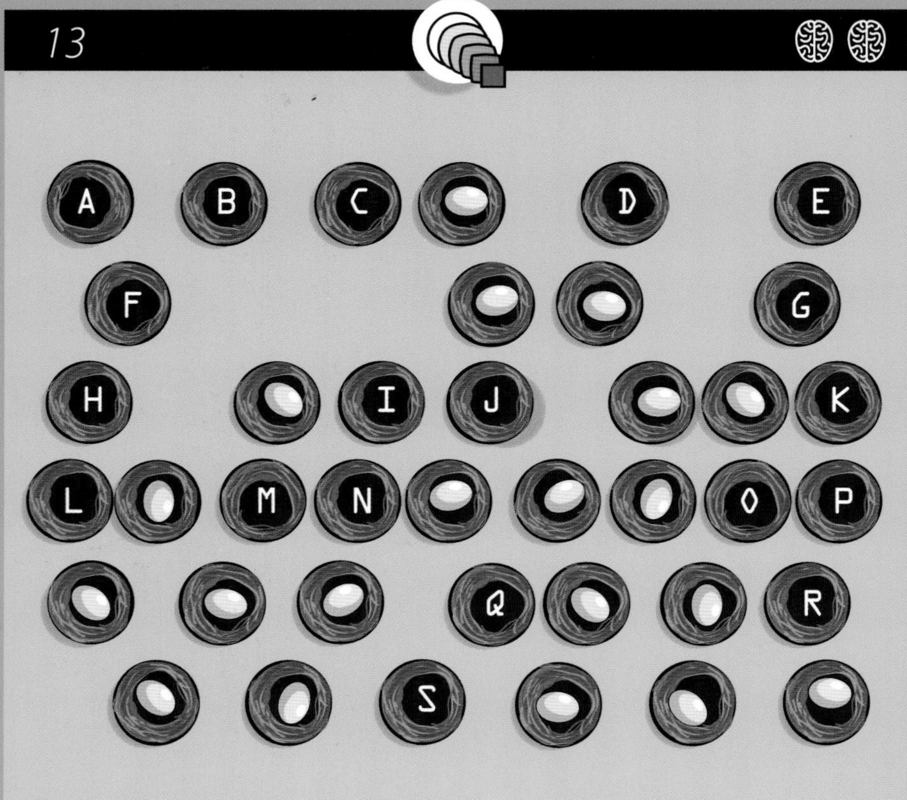

In which nest (A–S) will the bird lay the next egg?

BrainSnack® Puzzles

14

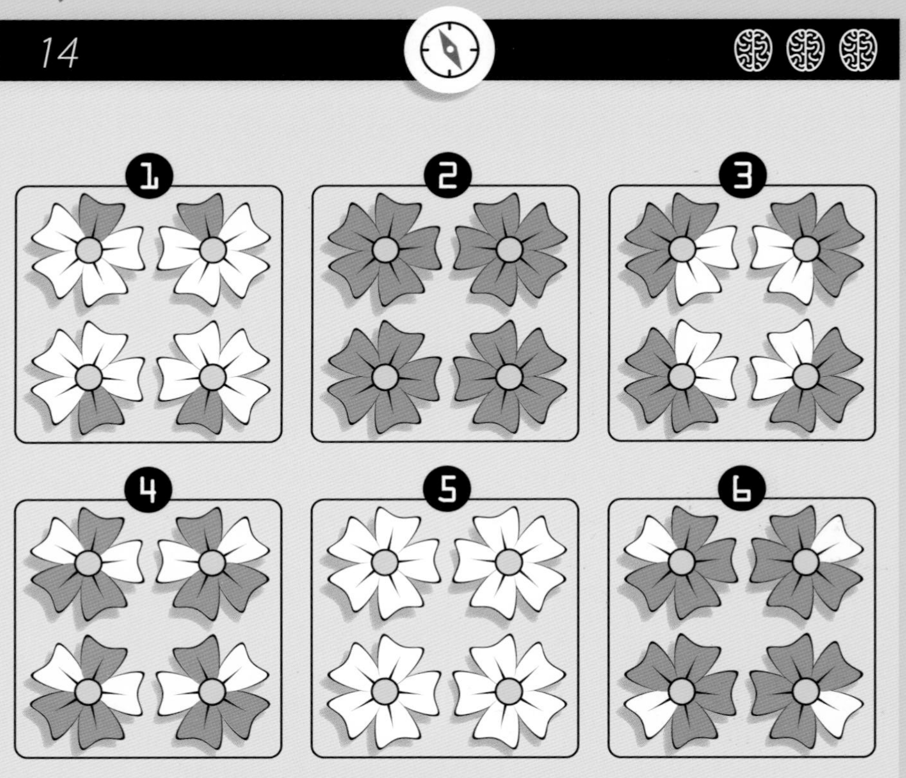

Which set of flowers (1–6) does not belong?

BrainSnack® Puzzles

15  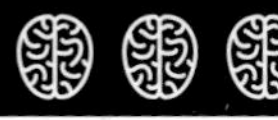

100101001011010100001010010
010010100100010010111010101
010100101010100101010101010
101010101000101010001011111
000101001001010010101010010

▶010010110100001000010000010
111010101010010101001010101
010101010101000101000101111
001000100101010010◀

▶001011000000000001011101010
100101001010101010101010100
010001011100001010???◀

The top binary string of five lines was simplified twice by converting the same binary code into one digit.

Which code should replace the question marks?

BrainSnack® Puzzles

16

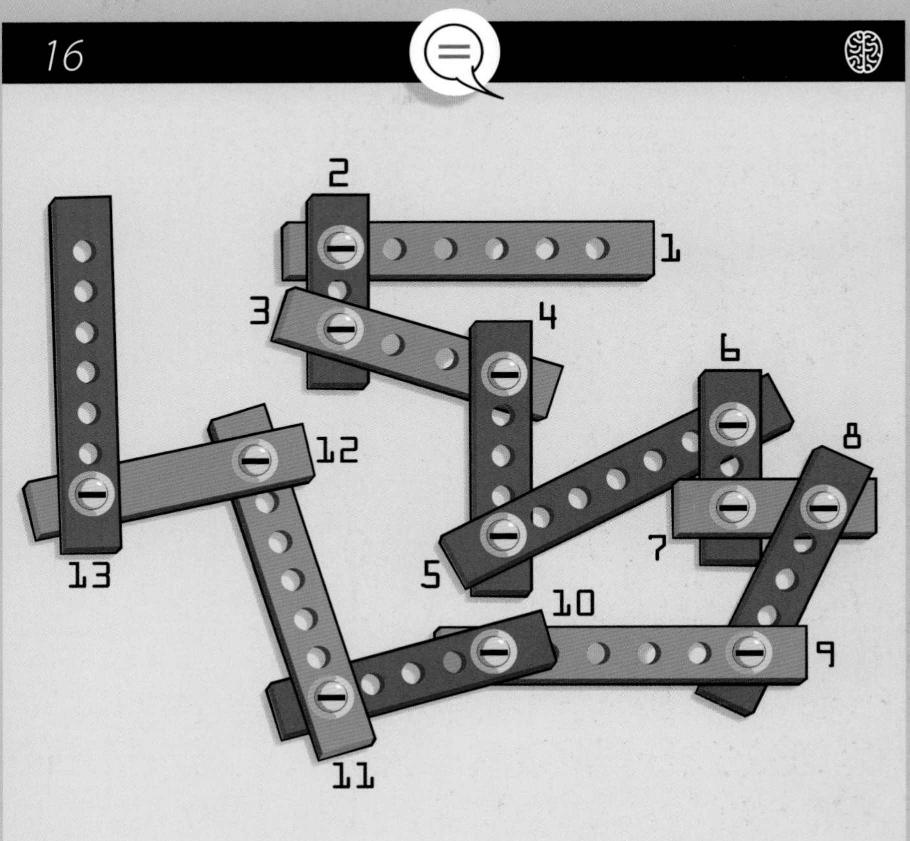

Which lath (1–13) does not belong?

BrainSnack® Puzzles

17  

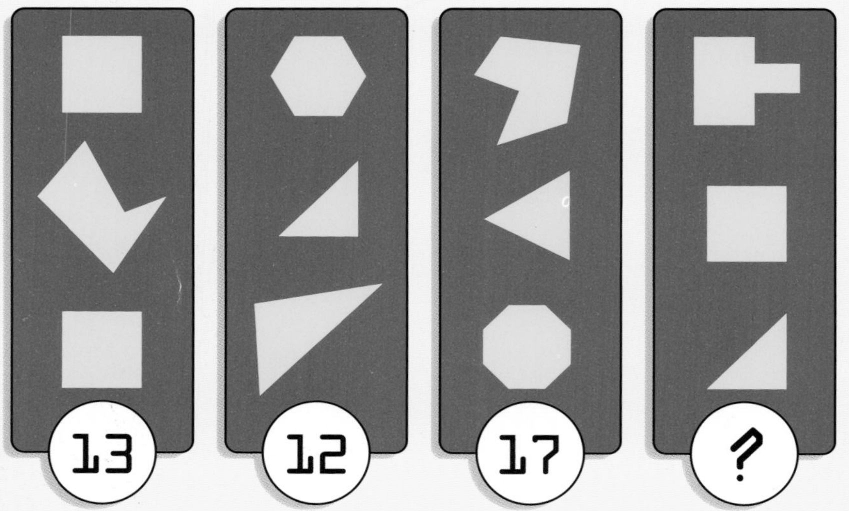

What number should replace the question mark?

BrainSnack® Puzzles

18

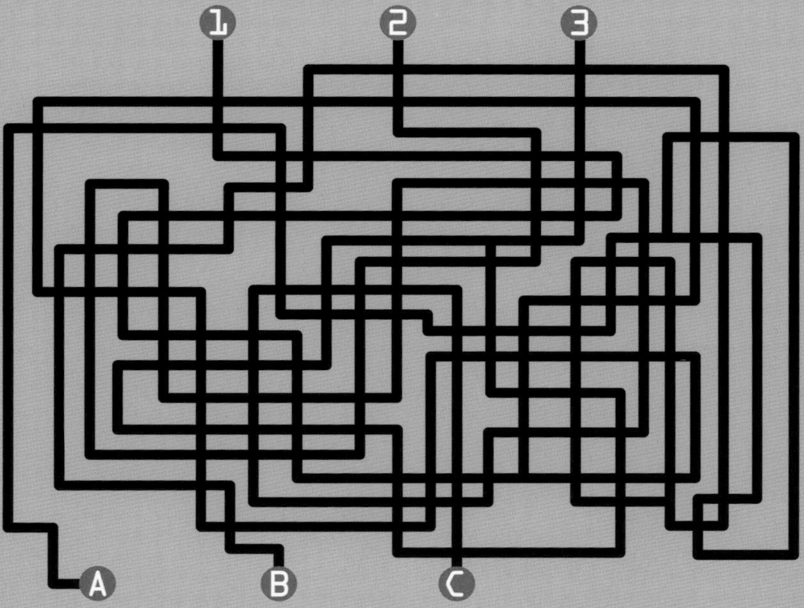

1 2 3

A B C

Which number can you connect with a letter? You can only go straight, so do not change directions at an intersection.

Only use your eyes. Don't use a pointer.

BrainSnack® Puzzles

19

Which honeybee (1–6) doesn't belong to the same family as all the others?

BrainSnack® Puzzles

Surfboards with the same color sail are on the same team.

Which surfboard has the wrong contest number?
Answer like this: 151617.

BrainSnack® Puzzles

21

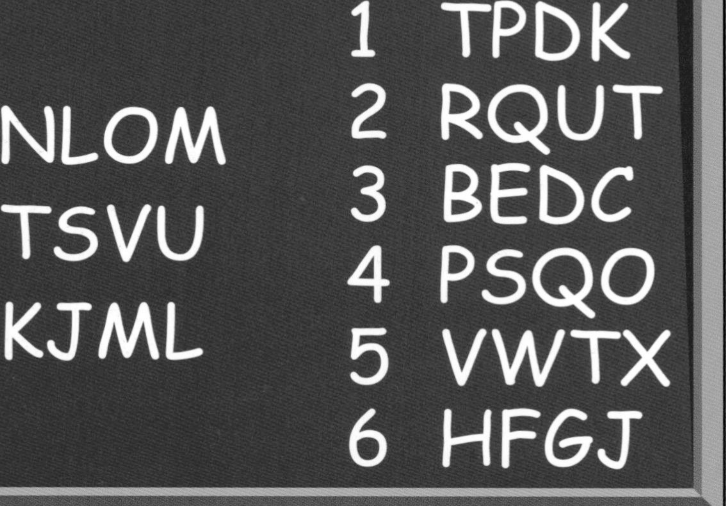

Which row (1–6) fits in the series on the left side?

BrainSnack® Puzzles

22

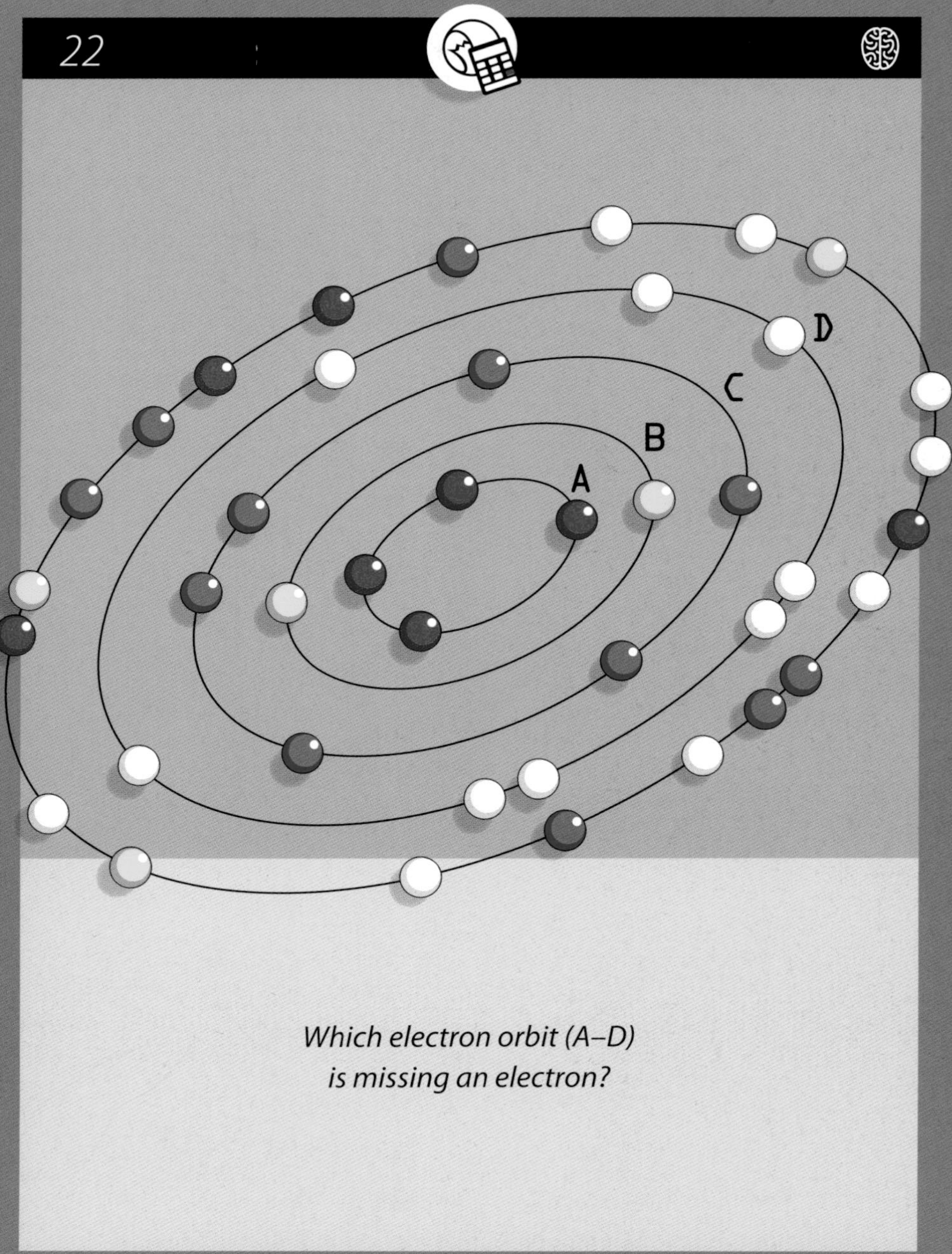

Which electron orbit (A–D) is missing an electron?

23

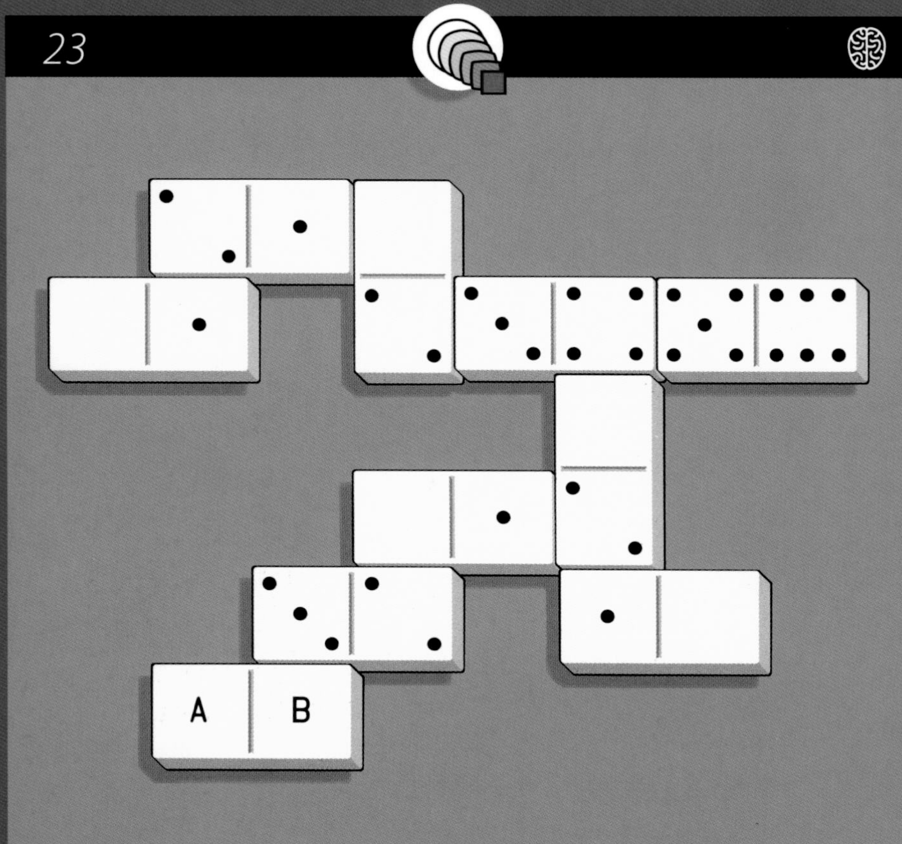

How many spots should replace the A and the B on the bottom domino?

BrainSnack® Puzzles

24

quis nostu dexerci tations ullamcorper suscipit lobortis nisl ut aliquip lexea commo doconsequa. Duis autem vel eum iriure dolor in hendrerit in vulputate velit esse molestie consequat, vel illum dolore eu feugiat nulla facilisis at vero eros et accumsan et.

Which word should be underlined?

BrainSnack® Puzzles

25

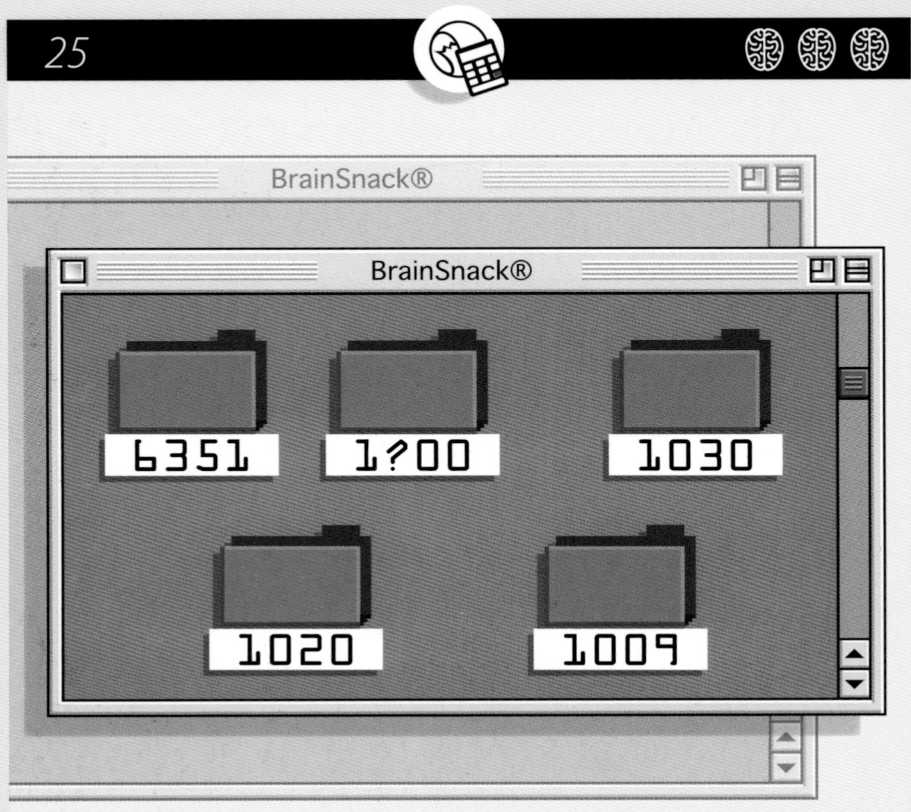

What digit should replace the question mark?

BrainSnack® Puzzles

26

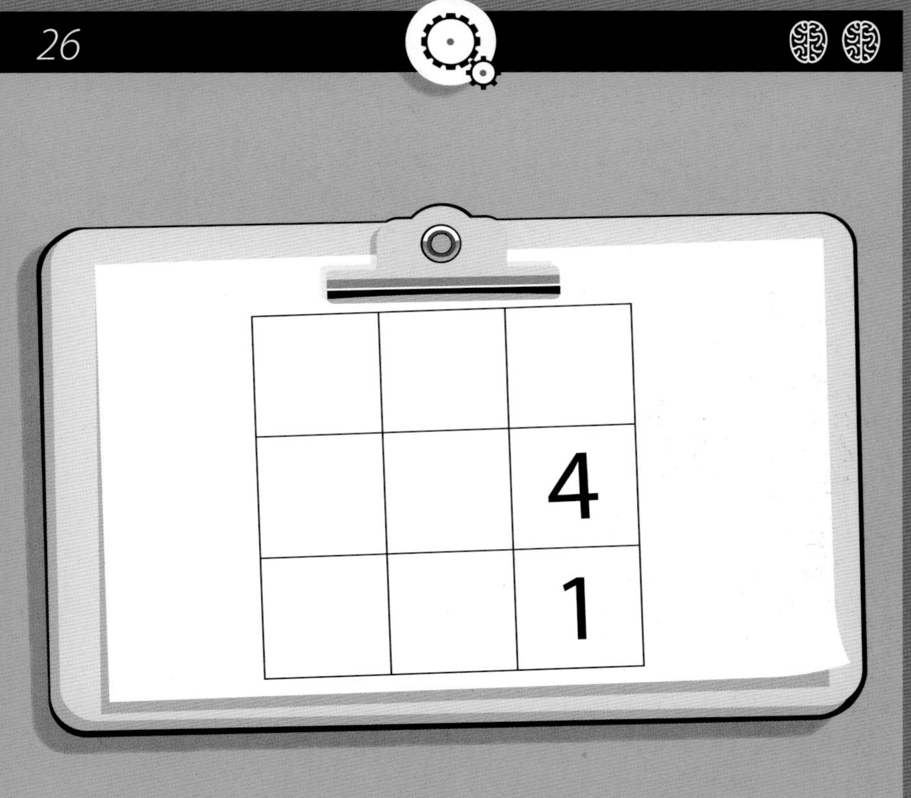

Fill in the chart with the numbers 1–9 so that the three-digit number on the top row equals 1/2 of the middle number and 1/3 of the bottom number.

BrainSnack® Puzzles

Which vowel should replace the question mark?

BrainSnack® Puzzles

28

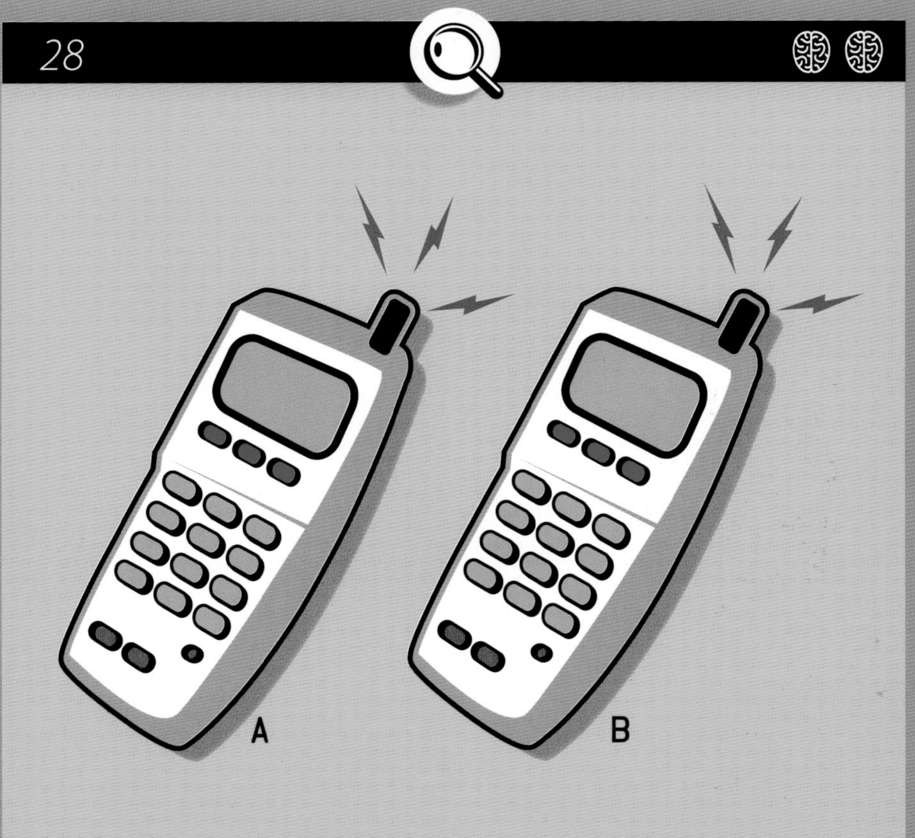

Spot the difference between images A and B.

BrainSnack® Puzzles

29

187 194 198 206 212 ?

How far will the skier jump on his last attempt?

BrainSnack® Puzzles

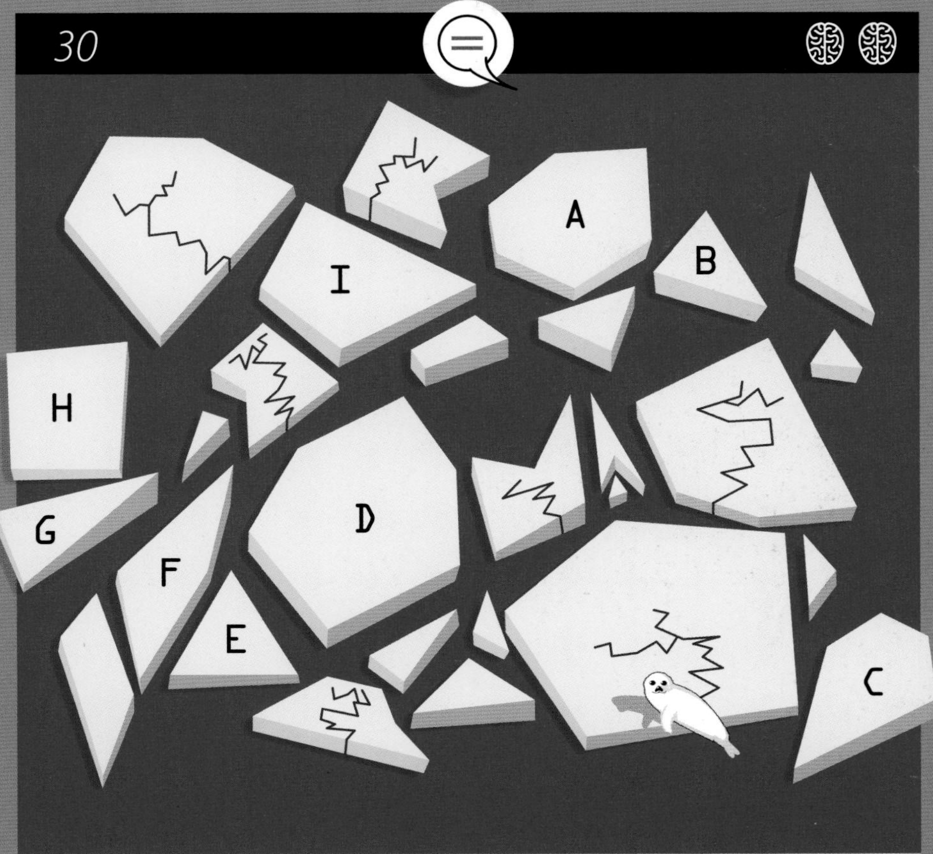

The young seal has rested on all the broken pieces of ice.

There is one more piece of ice he can rest on. Which piece (A–I) is it?

BrainSnack® Puzzles

31

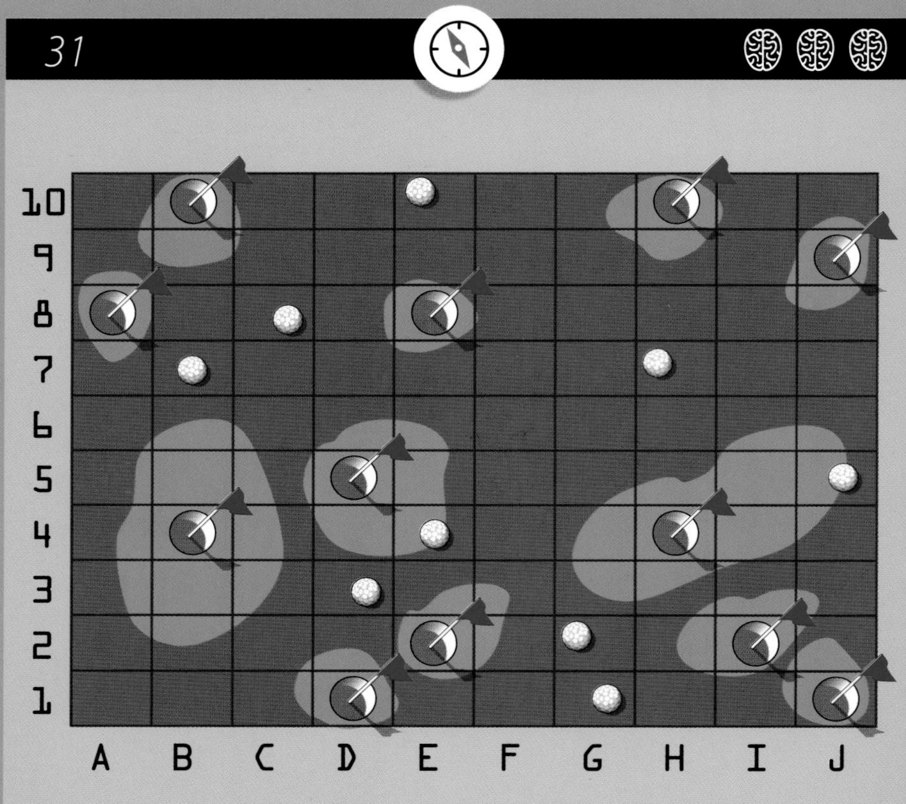

Which square is missing its golf ball?
Answer like this: C3.

BrainSnack® Puzzles

32

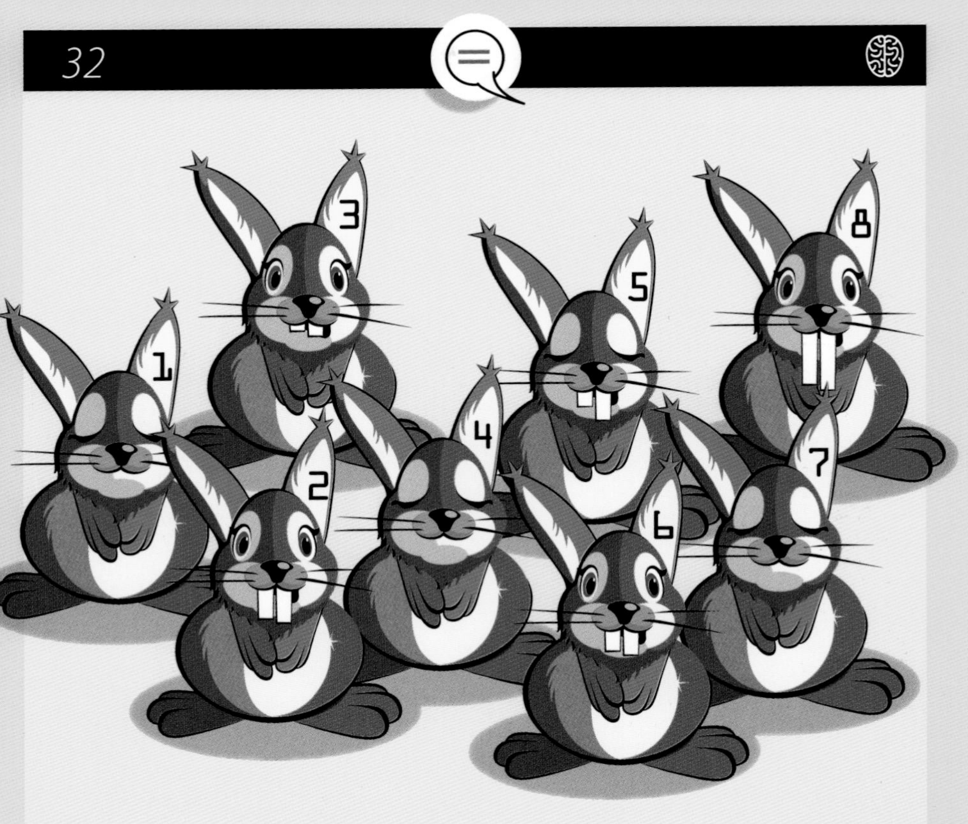

Which rabbit (1–8) does not belong?

33

How many blue stars are missing on the last firework?

BrainSnack® Puzzles

34

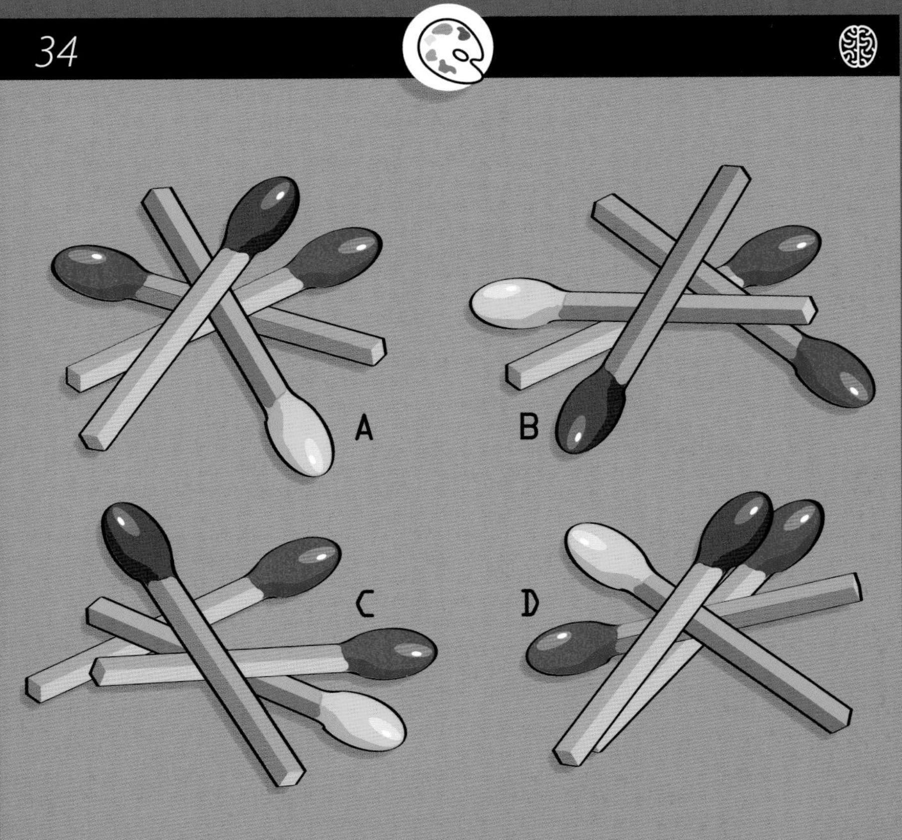

Which group of matches (A–D) doesn't belong in this series?

BrainSnack® Puzzles

35

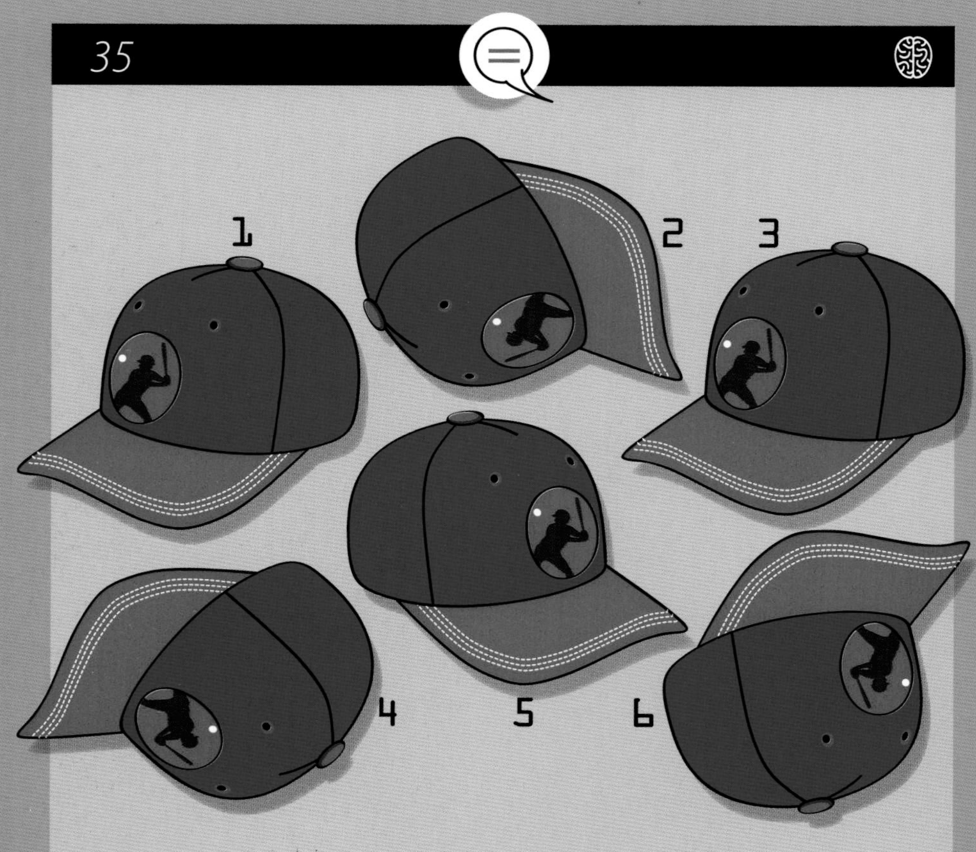

Which baseball cap (1–6) does not belong?

BrainSnack® Puzzles

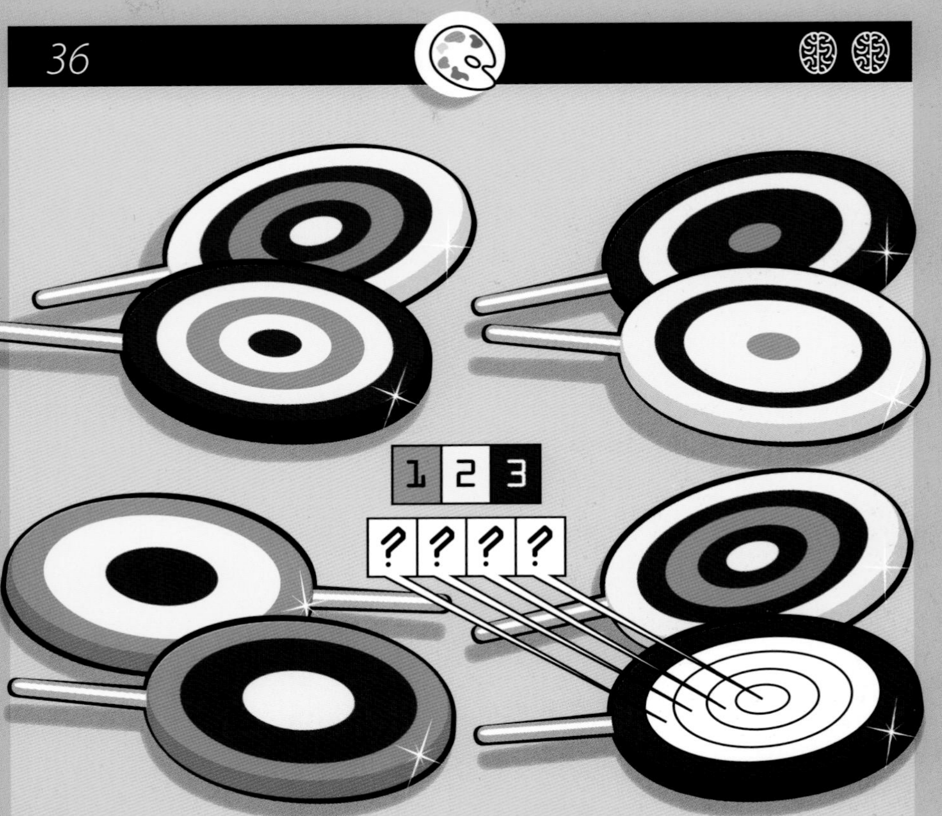

Give the color code of the inner rings of the lollipop.

Answer like this: 1223

BrainSnack® Puzzles

37  

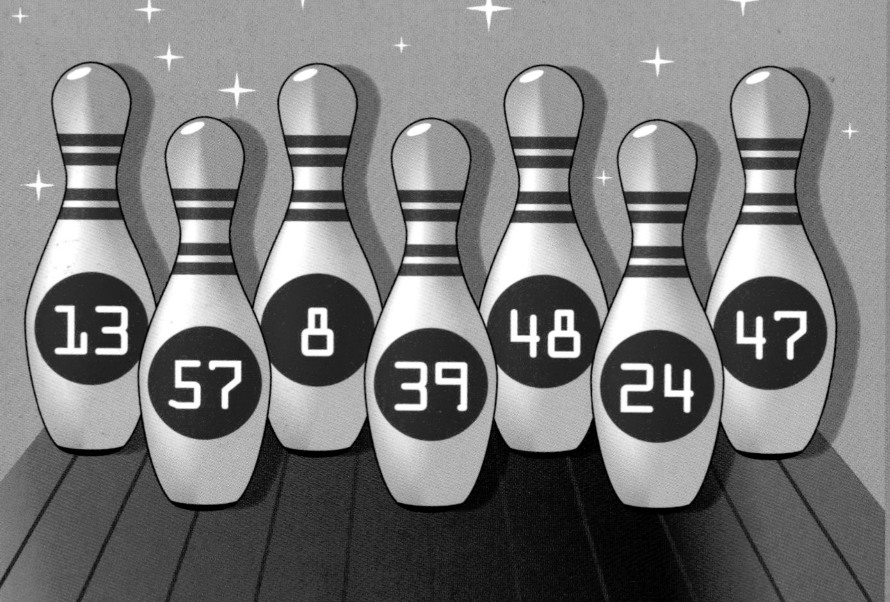

Which pins must be knocked over to score exactly one hundred points?

BrainSnack® Puzzles

38

What number should replace the question mark?

BrainSnack® Puzzles

39

20

The sum of five different numbers equals twenty. What is the biggest possible number you can use in this sum?

BrainSnack® Puzzles

40

What four digits complete this account number?

BrainSnack® Puzzles

41

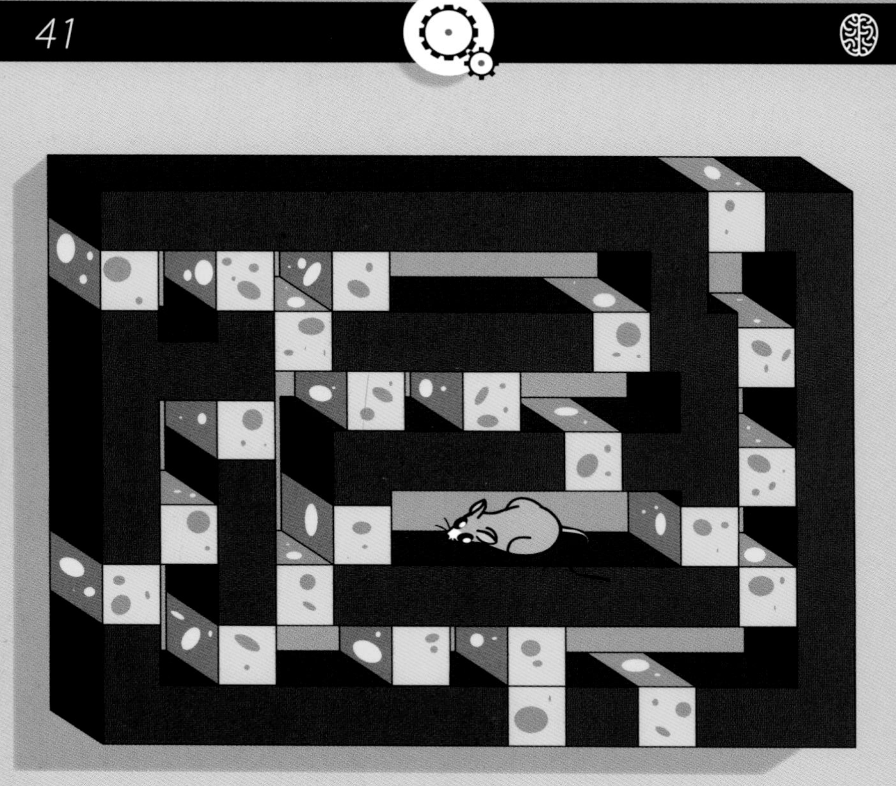

What is the minimum number of cheese cubes that the mouse has to eat to escape from the maze?

BrainSnack® Puzzles

42

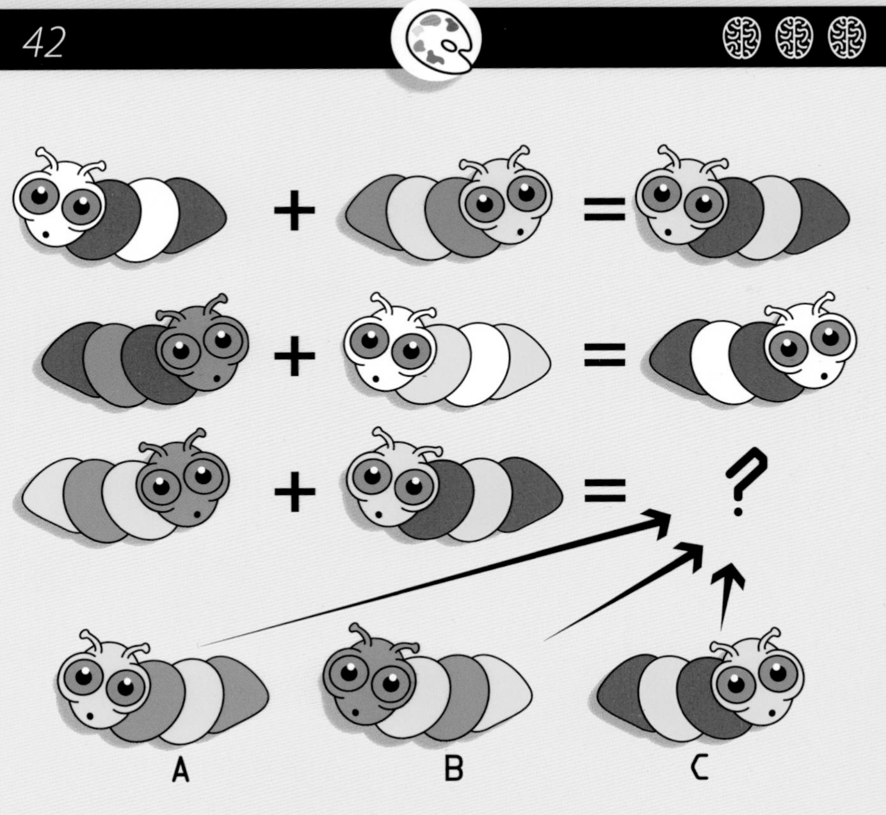

Just like in genetics,
dominant colors are at play here.

Which worm (A–C) should replace the question mark?

BrainSnack® Puzzles

43

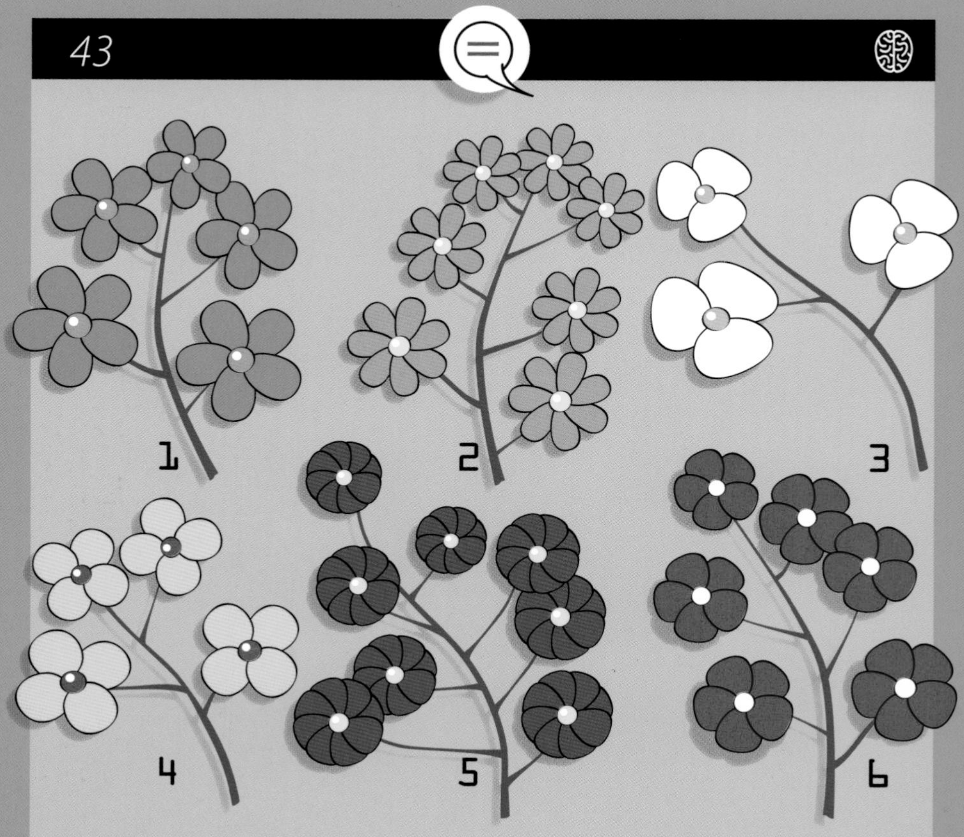

Which flower branch (1–6) does not belong?

BrainSnack® Puzzles

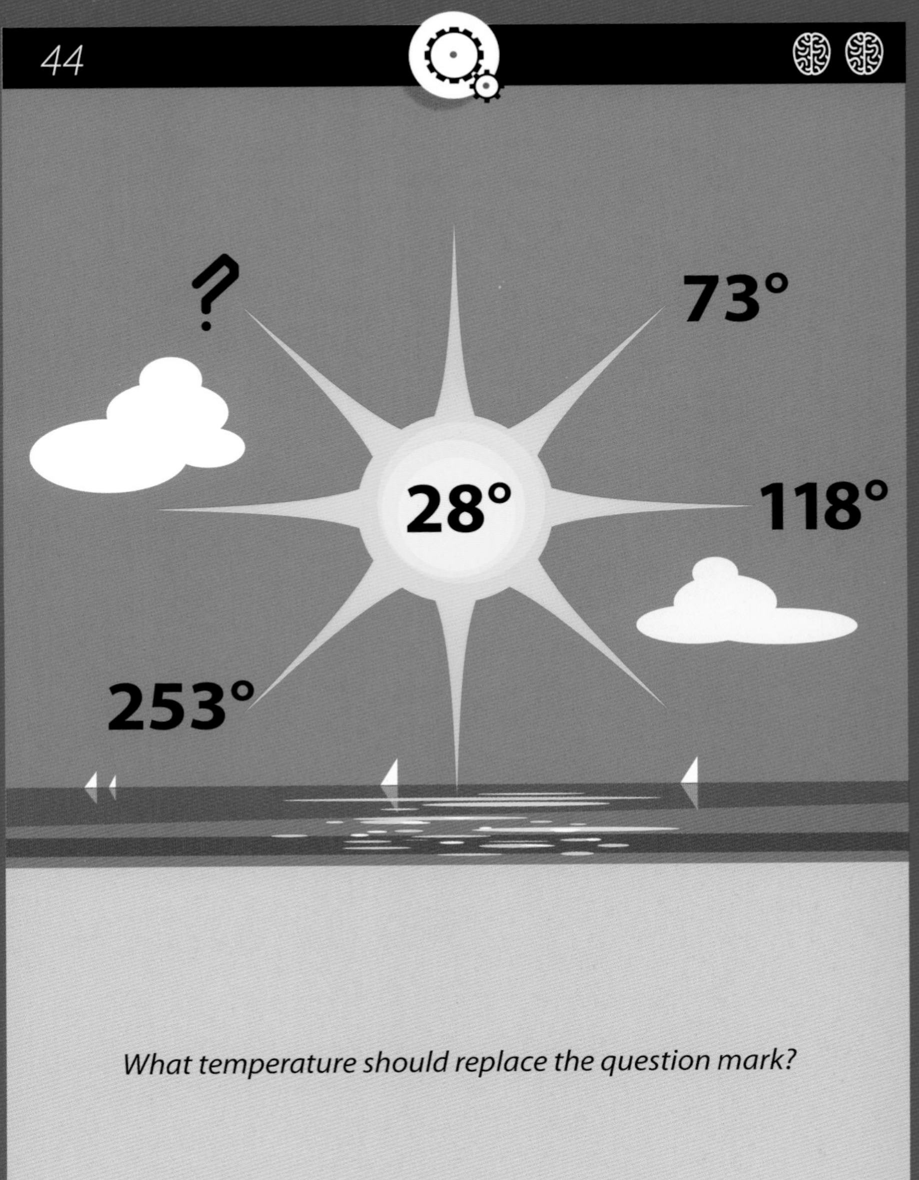

What temperature should replace the question mark?

BrainSnack® Puzzles

45

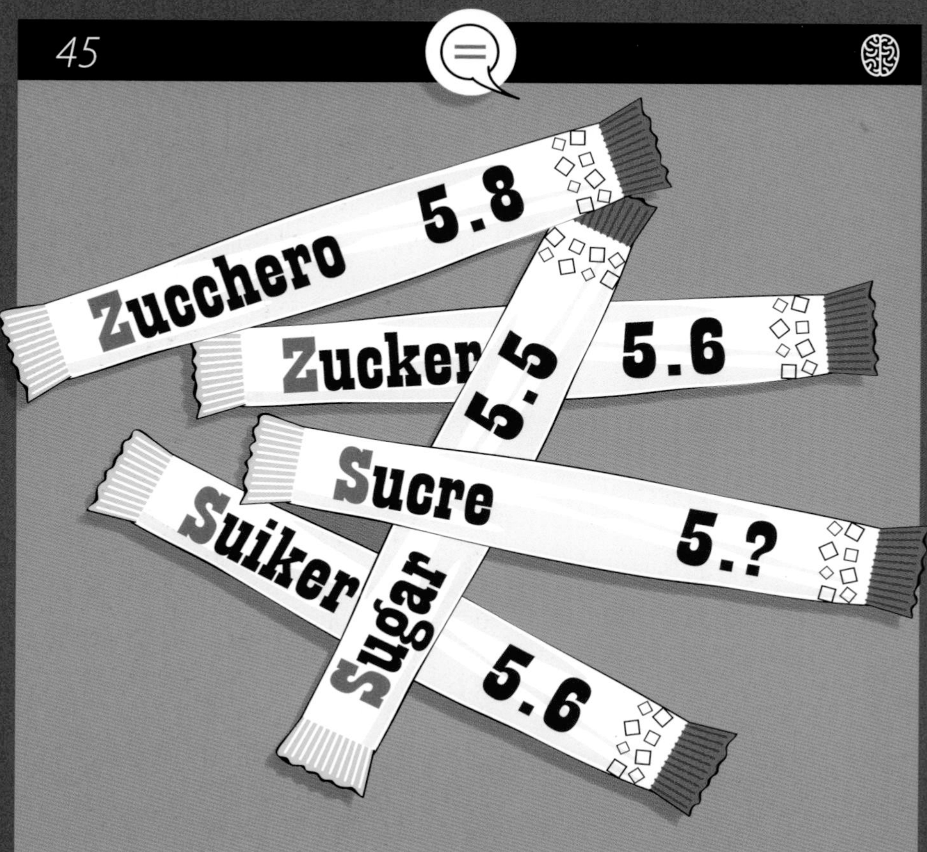

The weight of sugar is different per country.

What number should replace the question mark on the 'Sucre' packet?

BrainSnack® Puzzles

46

Which yogurt container (1–6) does not belong?

BrainSnack® Puzzles

At what point (1–7) on the dotted line will the hat be as tall as it is wide?

BrainSnack® Puzzles

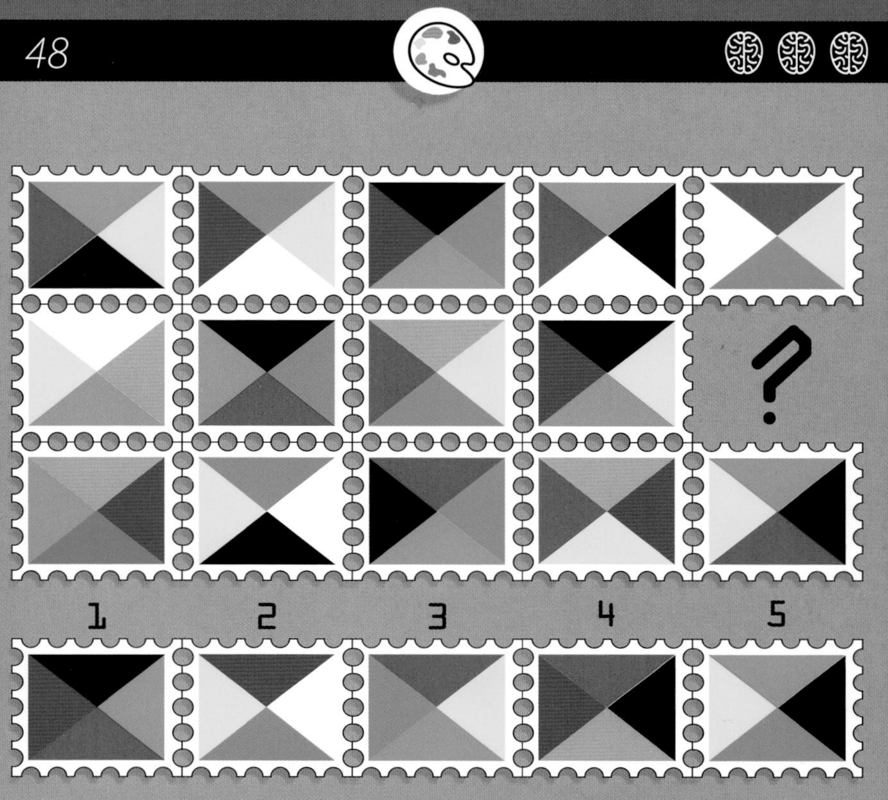

Which stamp (1–5) should replace the question mark?

BrainSnack® Puzzles

49

What number should replace the question mark?

BrainSnack® Puzzles

50

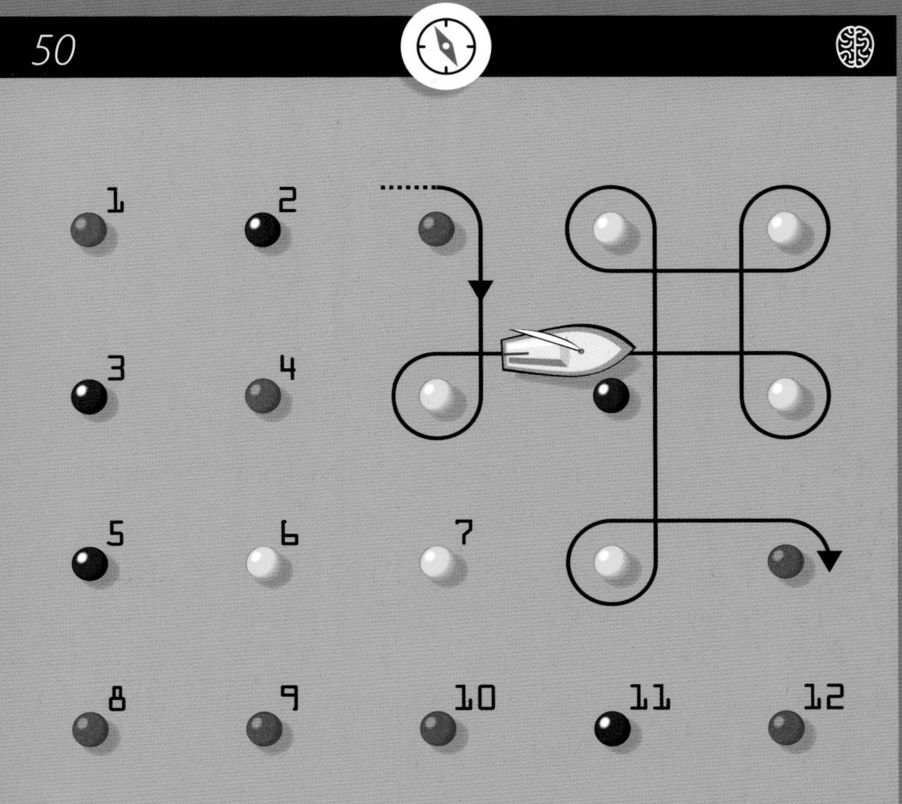

Which buoy (1–12) will the skipper never reach if he continues his journey using the same logic?

BrainSnack® Puzzles

51

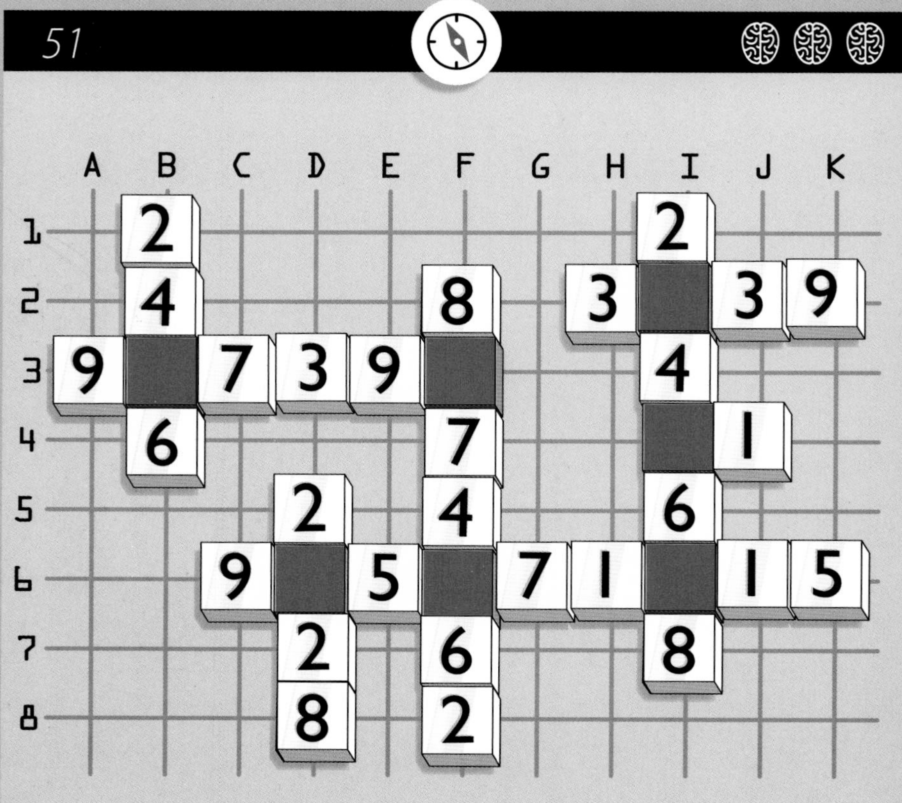

Which cube is in the wrong location?
Answer like this: 3D.

BrainSnack® Puzzles

52

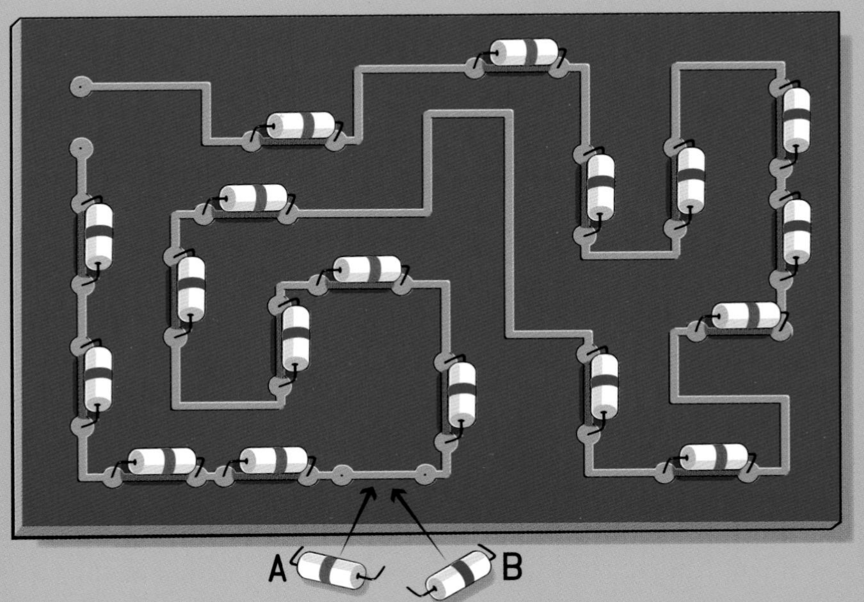

Which resistor (A–B) is missing here?

BrainSnack® Puzzles

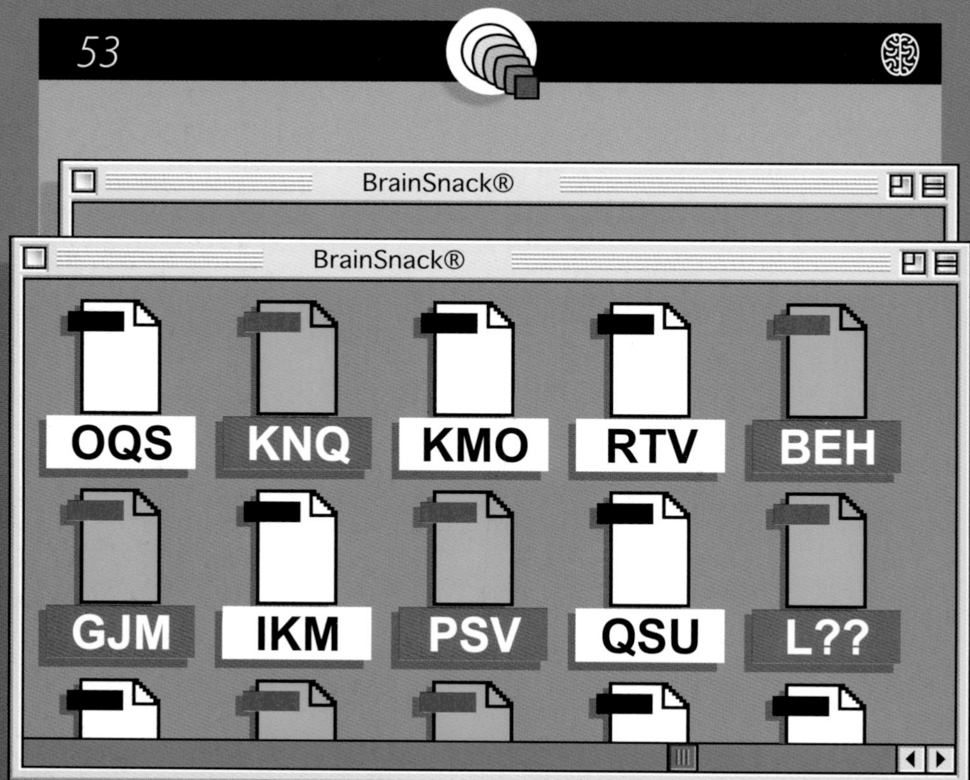

What two letters should replace the question marks?

BrainSnack® Puzzles

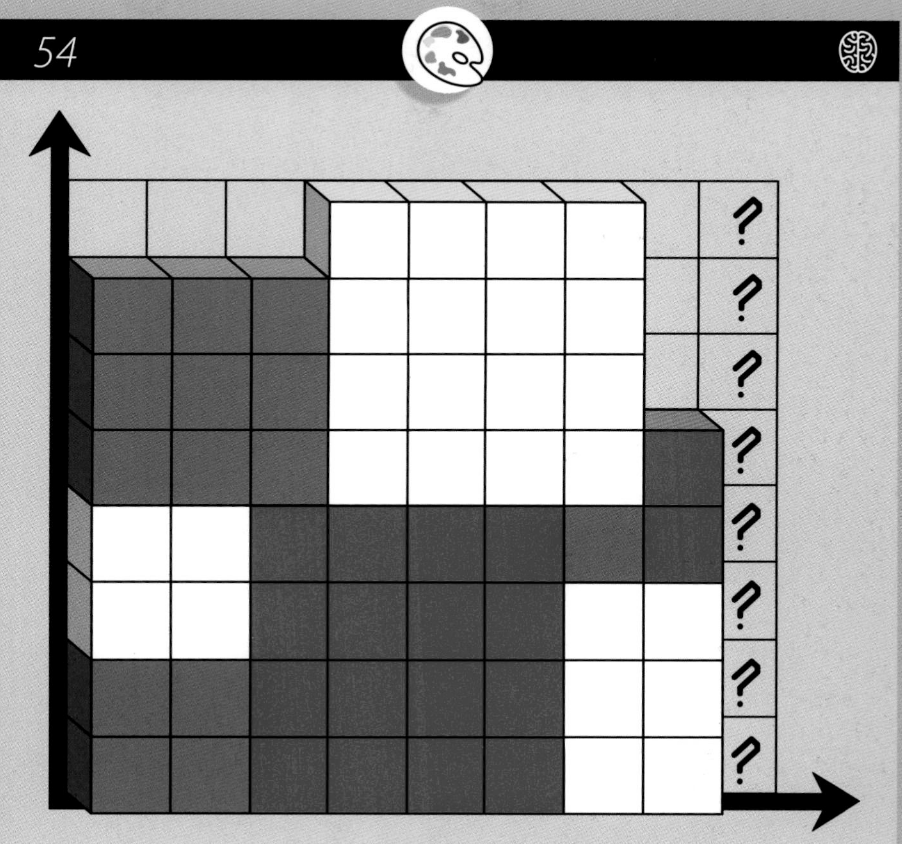

How many blocks are missing in the last column of this chart?

BrainSnack® Puzzles

55  

1	2	3
x j t l h w u	b v q t s d o	x p l t h b c

4	5	
c k s m i b o	k n p e q s z	a v r d e j l

Which square (1–5) contains none of the letters that appear in the last square?

BrainSnack® Puzzles

56

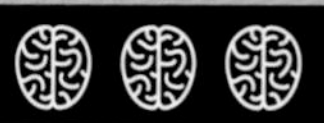

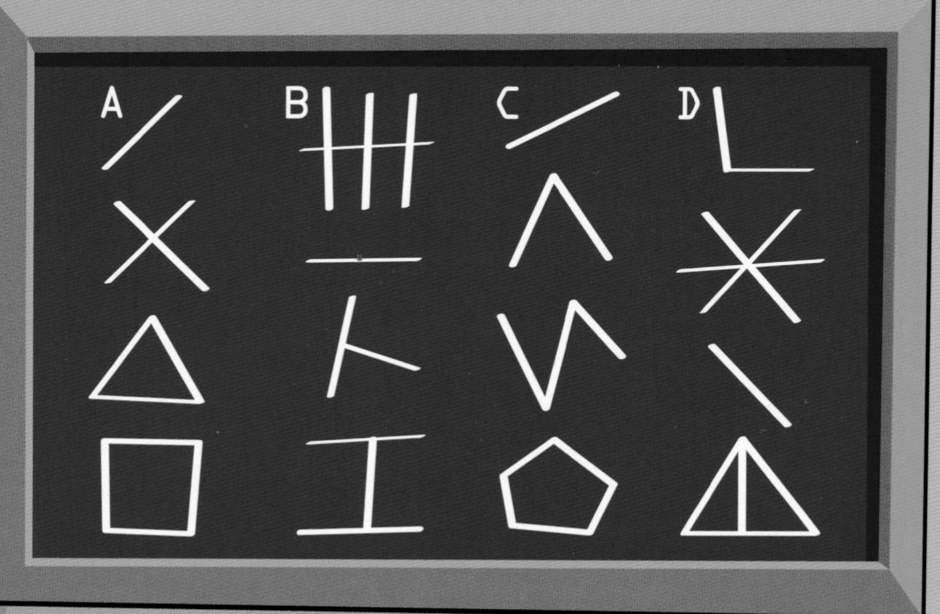

Which group of symbols (A–D) is unlike all the other groups?

BrainSnack® Puzzles

57

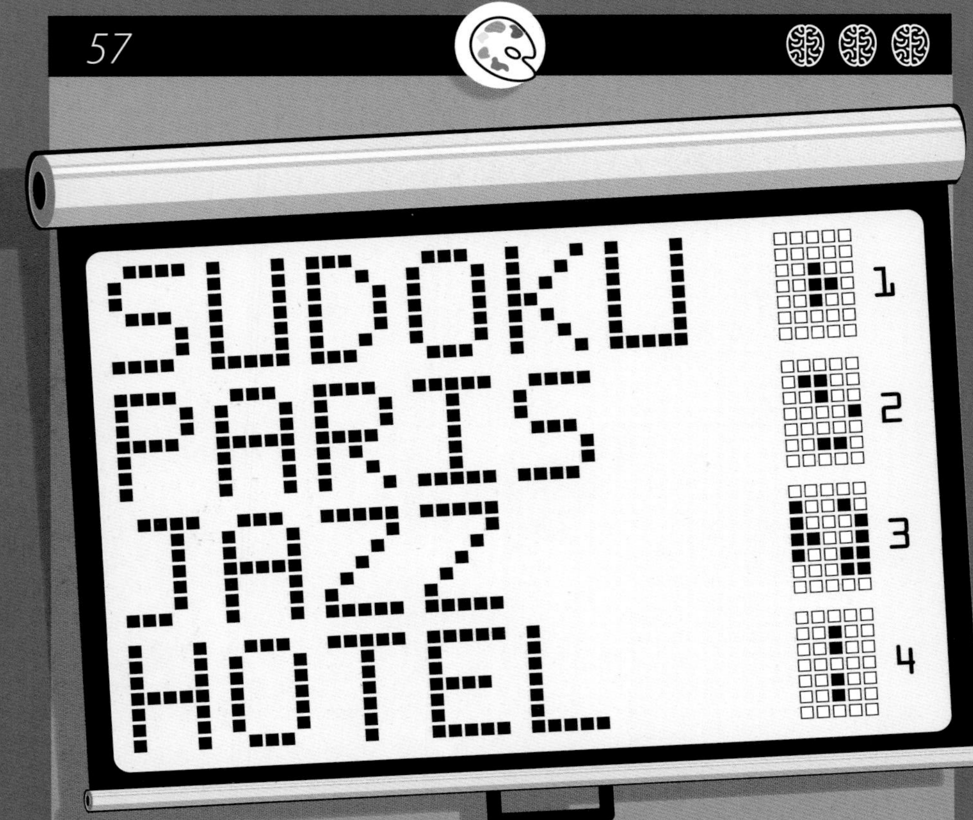

To the right of each word is a symbol that is related to the letters in the word.

Which symbol (1–4) is wrong?

BrainSnack® Puzzles

58

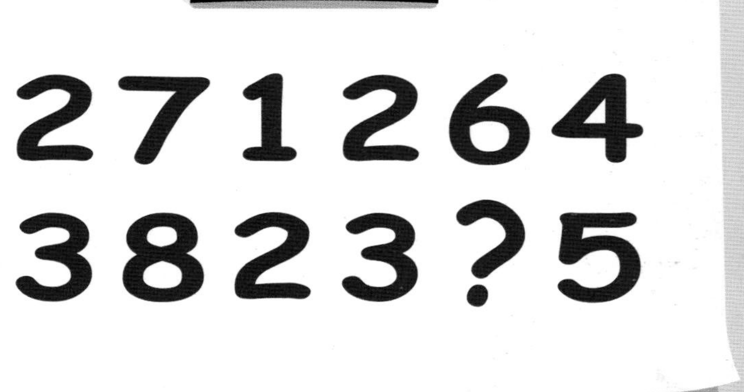

What number should replace the question mark?

BrainSnack® Puzzles

59

?

How many gold coins will the archaeologist find in the zone with a question mark?

BrainSnack® Puzzles

Which star (1–12) doesn't belong in the sky?

BrainSnack® Puzzles

61

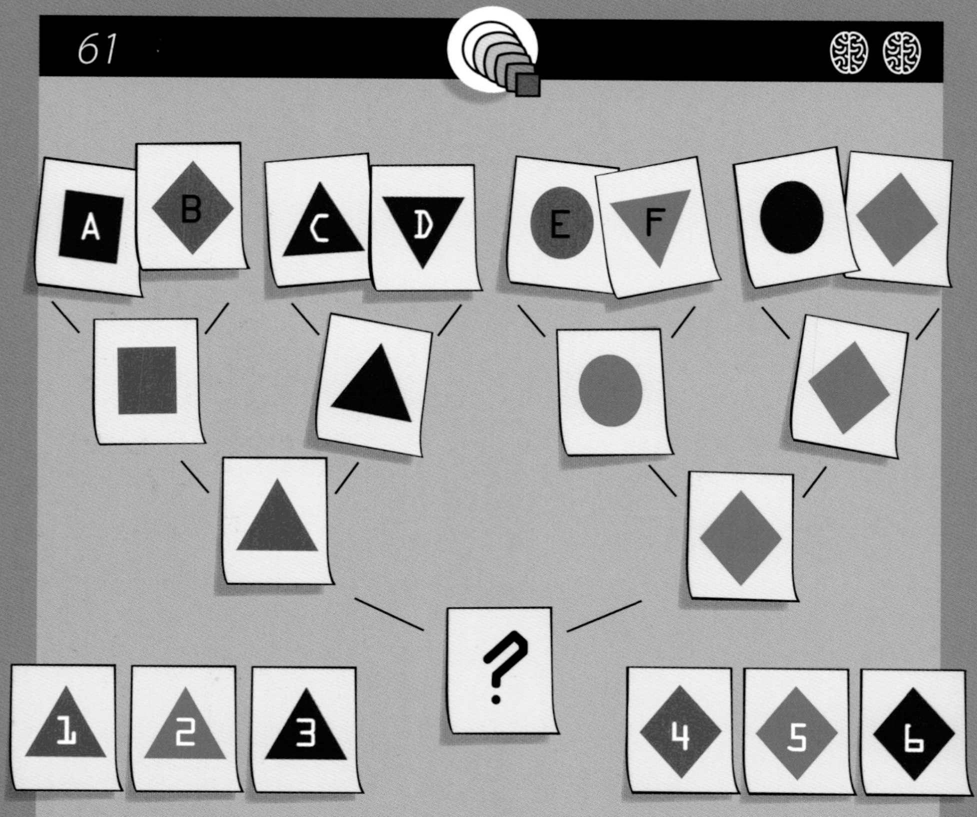

Which shape (1–6) will dominate in the end?

BrainSnack® Puzzles

62  

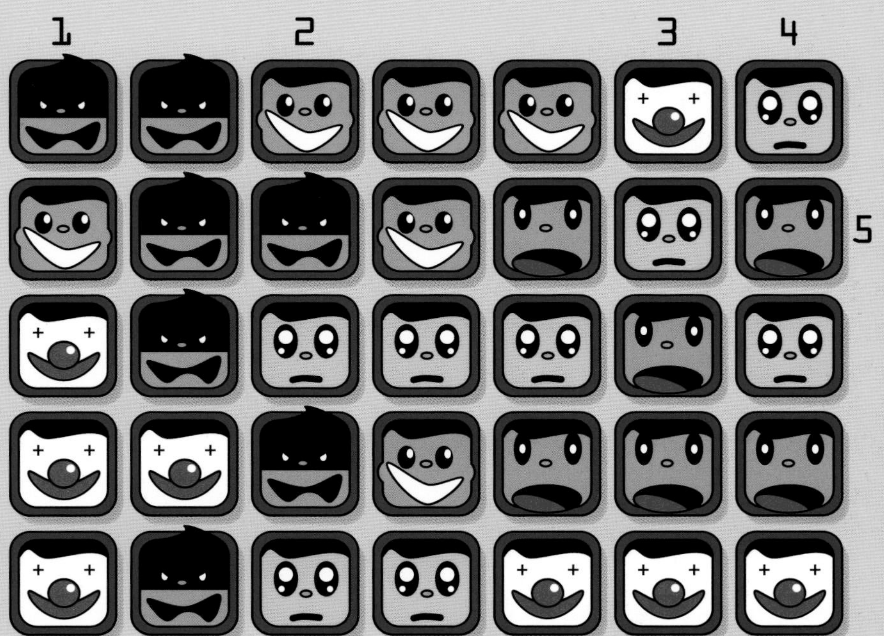

Which figure (1–5) do you see most frequently three times in a row in all directions?

Words & Letters

What is the meaning of the first two letters of the Phoenician alphabet from which our word 'alphabet' derives?

Solution on page 288

Teach

Learn

The Oxford English Dictionary contains full entries for 171,476 words in current use.

Our brain is the most extraordinary form of organized matter in our solar system, and it is also able to think that up and remember it.

Thinking and remembering are an essential process in everything that we do, even if we are not always aware of this. A lot of that thinking and remembering uses words.

With word puzzles, your memory will have to recall words and look for links in meaning and shape. If you want to recall a word, the brain must link the meaning of that word to as many other words with a similar meaning as possible. Your lexical memory is responsible for remembering the shape—the spelling—of the word.

If you do not recall the meaning of the word taxidermist, your intrinsic memory has failed you. If you no longer know the name of someone who stuffs animals, then your lexical memory has failed you and you say . . .

"It is on the tip of my tongue."

Words & Letters

63 LETTER BLOCS

1 *ATHLETICS*

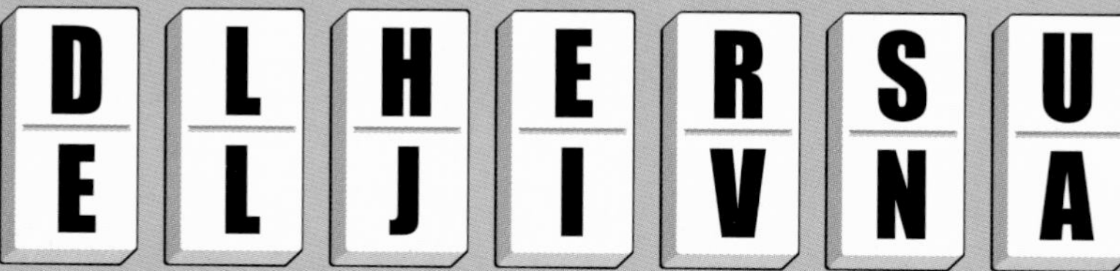

2 *MARRIAGE*

3 *DANCE*

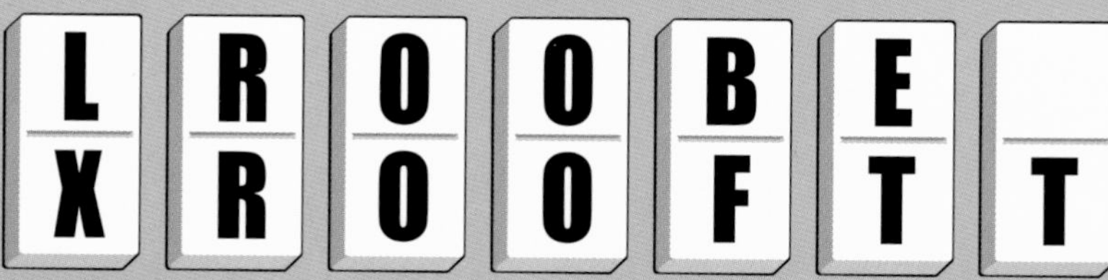

Move the letter blocks around so that words are formed on top and below that you can associate with the subject.

Words & Letters

64 ONE LETTER LESS OR MORE

1 *ATLANTIC*	*– N*	C									
2 *CAMPAIGN*	*– A*		A								
3 *HAIRLINE*	*– I*			H							
4 *BERMUDAS*	*+ T*				M						
5 *BACHELOR*	*– B*					E					
6 *EARPHONE*	*+ C*						R				
7 *INCREASE*	*– A*							E			

Each word on the right side contains the letters in the word at the left side, plus or minus the letter in the middle.

One letter is already in the right place.

Words & Letters

65 HOLIDAYS

1 *TROLLEY (quality lodging)*

	U	X	U				H				

2 *ONE LOBSTER (ice made from fruit juice)*

		M									

3 *WIT MISS (tight fitting garment)*

					U		

4 *CREAM (record memories)*

					A

5 *ICE VIRUS (go with the flow)*

R							R				E

Use the letters above each grid to form the word that is described in parentheses.

One or more extra letters are already in the right place.

Words & Letters

66 DOODLE PUZZLES

1 ....

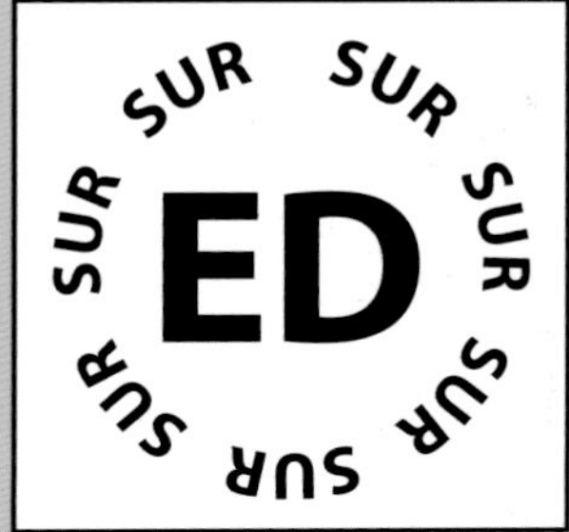

2 ....

WEIGHT
WEIGHT
WEIGHT
WEIGHT
WEIGHT

3 ....

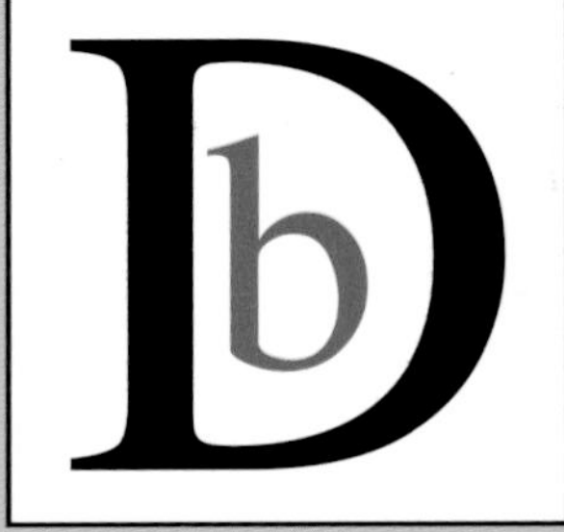

3 ....

A doodle puzzle is a combination of images, letters, and numbers that indicate a word or concept. If you cannot solve a doodle puzzle, do not look at the answer right away. Try to solve it later or tomorrow. When you know the answer, study the exercise and the solution to remember the structure and connections of the puzzle forever. Afterward it will be easy to solve doodle puzzles and explain them to your friends. This will reinforce your comprehension.

Words & Letters

67 BRAINSNACK

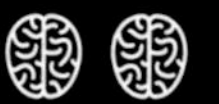

A	T	U	L	K	Q
B	N	C	D	V	R
E	M	I	S	W	H
F	Y	X	O	P	G

Cross off all pairs of letters that satisfy a certain logic. Which two letters are left over?

Words & Letters

68 WORD ASSOCIATIONS

1
chocolate . . .
....
....
....
....
....
. . . stress

2
shepherd . . .
....
....
....
....
....
. . . winter

3
luck . . .
....
....
....
....
....
. . . prison

4
car . . .
....
....
....
....
....
. . . awake

5
naked . . .
....
....
....
....
....
. . . train

6
football . . .
....
....
....
....
....
. . . school

Connect the two words in five steps by producing a new word that is associated with the previous word.

Several solutions are possible. Example:
apple**–tree–pulp–paper–page–number–**phone

Words & Letters

69 DOODLE PUZZLES

1 ....

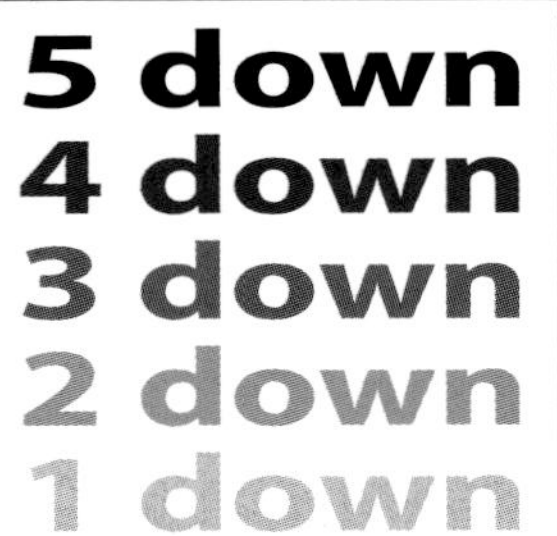

2 ....

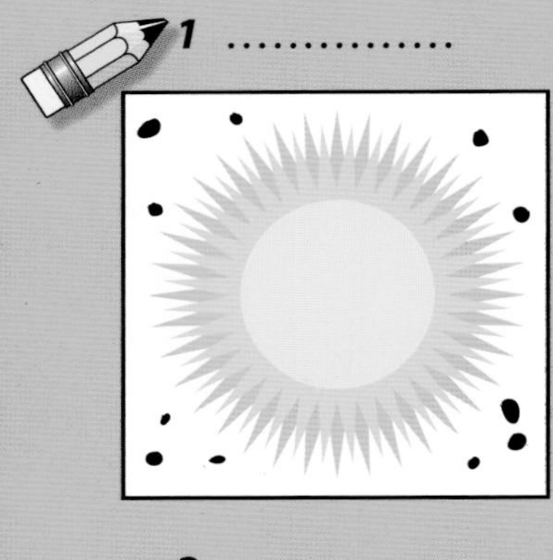

3 ....

4 ....

A doodle puzzle is a combination of images, letters, and numbers that indicate a word or concept. If you cannot solve a doodle puzzle, do not look at the answer right away. Try to solve it later or tomorrow.

Words & Letters

70 ONE LETTER LESS OR MORE

	Word		Answer
1	NEBRASKA	– A	B _ _ _ _ _ _
2	CANOEING	+ R	_ G _ _ _ _ _ _ _
3	GRENADES	+ E	_ _ N _ _ _ _ _ _
4	BASELINE	– E	_ _ _ B _ _ _
5	COMPLETE	– P	_ _ _ _ C _ _
6	FACELIFT	– E	_ _ _ _ _ C _
7	CHARISMA	+ R	_ _ _ _ _ _ I _ _

Each word on the right side contains the letters in the word at the left side, plus or minus the letter in the middle.

One letter is already in the right place.

Words & Letters

71 EUPHEMISMS

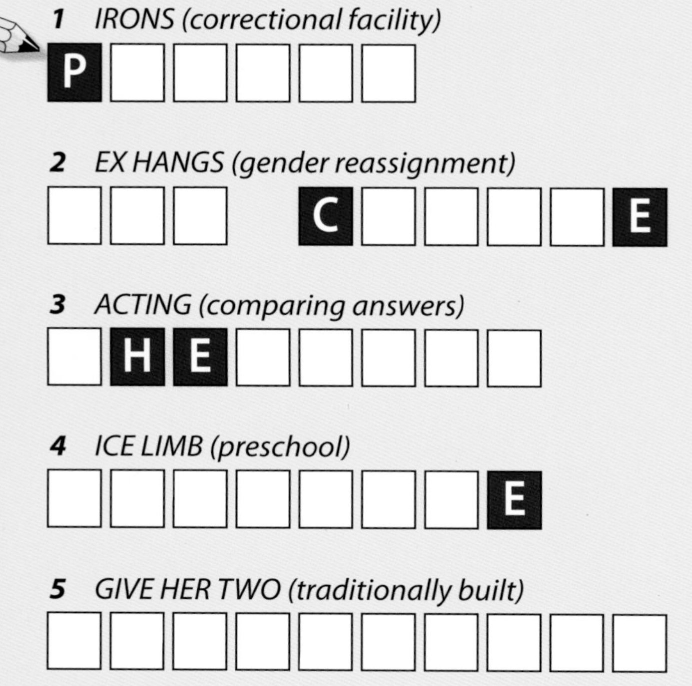

1 IRONS (correctional facility)
P
2 EX HANGS (gender reassignment)
C
E
3 ACTING (comparing answers)
H
E
4 ICE LIMB (preschool)
E
5 GIVE HER TWO (traditionally built)

Use the letters above each grid to form the word that is described by the euphamism in parentheses.

Form the word that is described with the euphemism between parentheses with the letters above each grid.

One or more extra letters are already in the right place.

Words & Letters

72 SWEET WORD PYRAMID

(1) *in the Christian era*
(2) *boy*
(3) *placed*
(4) *perfect*
(5) *dames*
(6) *misdirect*
(7) *philosophical theory that ideas are the only reality*

Each word in the pyramid has the letters in the word above it, plus a new letter.

Words & Letters

73 LETTER BLOCKS

1 *COURT*

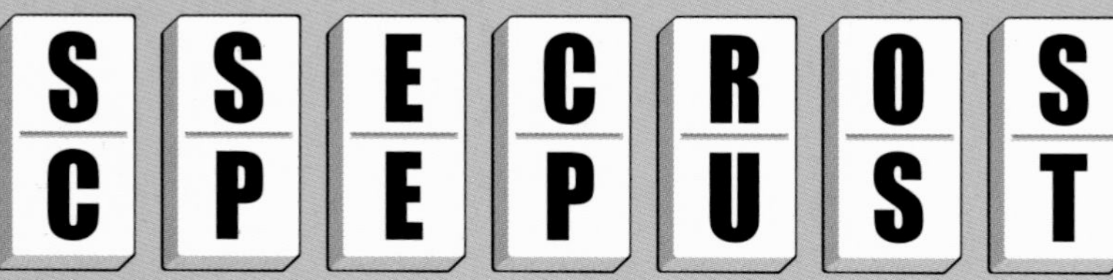

2 *MOTORCYCLE*

3 *EMOTIONS*

Move the letter blocks around so that words are formed on top and below that you can associate with the subject.

The letters were reversed on one block.

Words & Letters

74 ANIMAL SOUNDS

1 *APES* _ _ _ B _ _

2 *ELEPHANTS* _ _ _ _ _ _ T

3 *COCKS* _ _ _ _

4 *DOGS* _ _ _ _

5 *FLIES* _ U _ _

6 *HAWKS* _ _ _ _ A _

7 *PENGUINS* _ _ _ K

Fill in the appropriate verb that corresponds with the sound made by the animals.

One letter is already in the right place.

Words & Letters

75 ANIMAL SPECIES

1 BELLOW	_ _ _ L _
2 WHINE	_ _ _ _ _ _ T _ _ _
3 HISS	_ N _ _ _ _
4 GOBBLE	_ _ _ _ E _ _
5 TALK	_ _ _ _ O _ _
6 CROAK	_ _ _ G _
7 CRY	S _ _ _ _

Fill in the appropriate animal name that corresponds with the sound made by the animals.

One letter is already in the right place.

Words & Letters

76 PHOBIAS

1 GIN FLY (Aviophobia is fear of . . .)
2 DEEP (Tachophobia is fear of . . .)
S
3 OWN ME (Gynophobia is fear of . . .)
4 NESTING (Neophobia is fear of . . .)
W
H
5 GROW INK (Ergophobia is fear of . . .)

A phobia is an irrational or obsessive fear or anxiety, usually of or about something particular. For example, hippopotomonstrosesquipedaliophobia is the fear of long words.

Use the letters above each grid to form the word that completes the phobia described in parentheses.

One or more extra letters are already in the right place.

Words & Letters

77 ONE LETTER LESS OR MORE

1 *LOCATION*	*+ I*	C								
2 *BEDCOVER*	*– B*		O							
3 *DOORSTEP*	*– S*			R						
4 *COLLAPSE*	*– O*				L					
5 *KINGFISH*	*– K*					I				
6 *DIAMETER*	*+ G*						A			
7 *NAPOLEON*	*+ M*							A		

Each word on the right side contains the letters in the word at the left side, plus or minus the letter in the middle.

One letter is already in the right place.

Words & Letters

78 BRAINSNACK

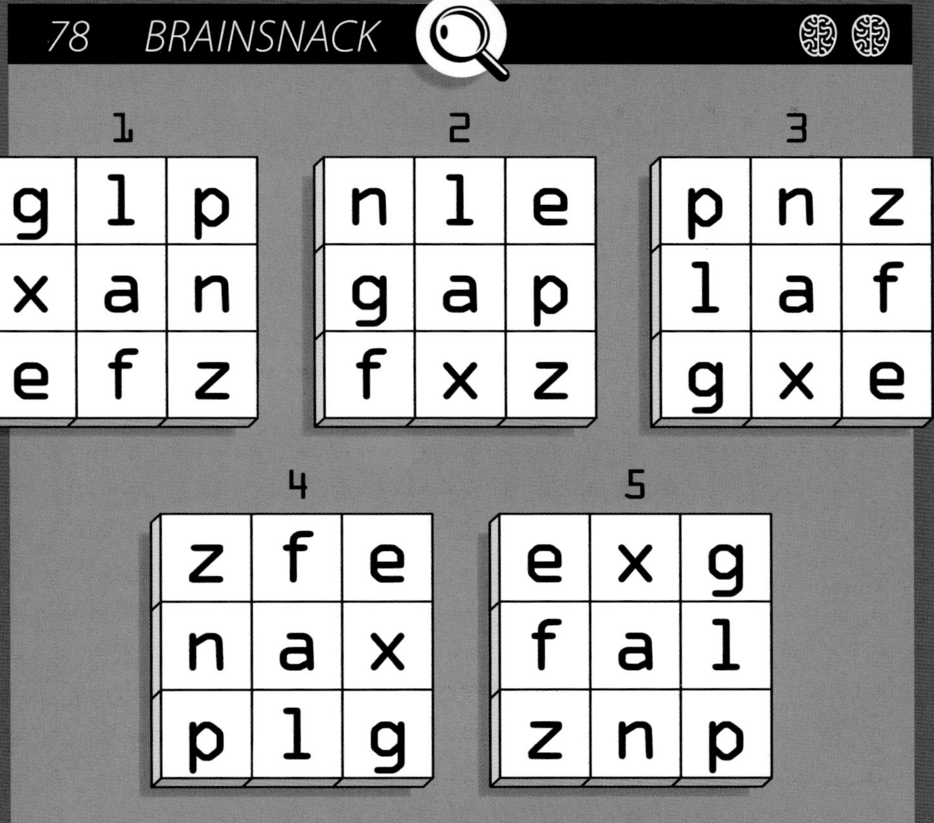

Which group of letters (1–5) does not belong?

Words & Letters

79 WORDPLAY

1 *LATIN COP (always joins the party)*

_ _ _ I _ _ _ I _ _

2 *ARTISTIC SPY (cannot be disturbed)*

_ _ _ _ H _ _ _ _ _ _ _

3 *CHUCK (realizes the worst is behind him)*

_ _ N _ H B A _ _

4 *AIR GAME (to some a word, to others a sentence)*

_ _ _ R _ _ _ _

5 *JUNE BUG (it will get you suspended)*

_ _ _ _ _ E _ _ M P

Use the letters above each grid to form the word that is described by the wordplay in parentheses.

One or more extra letters are already in the right place.

Words & Letters

80 SWEET WORD PYRAMID

Each word in the pyramid has the letters in the word above it, plus a new letter.

(1) musical note

(2) plump

(3) quick

(4) realities

(5) workmanships

(6) elements

(7) prediction

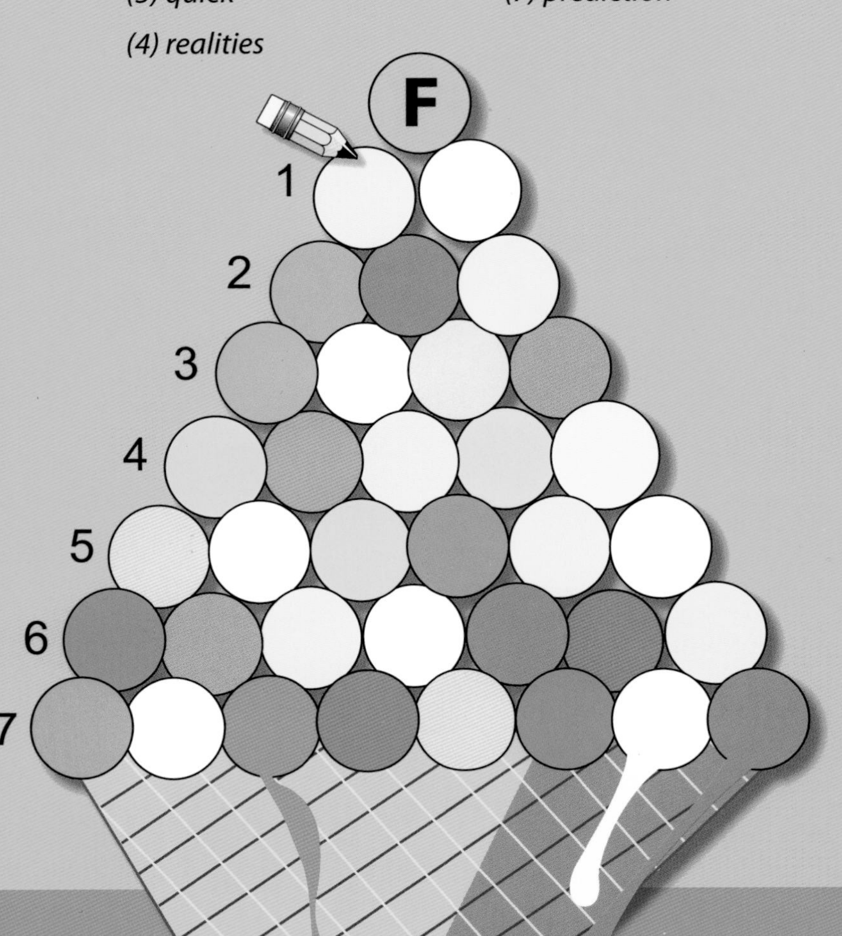

Words & Letters

81 ONE LETTER LESS OR MORE

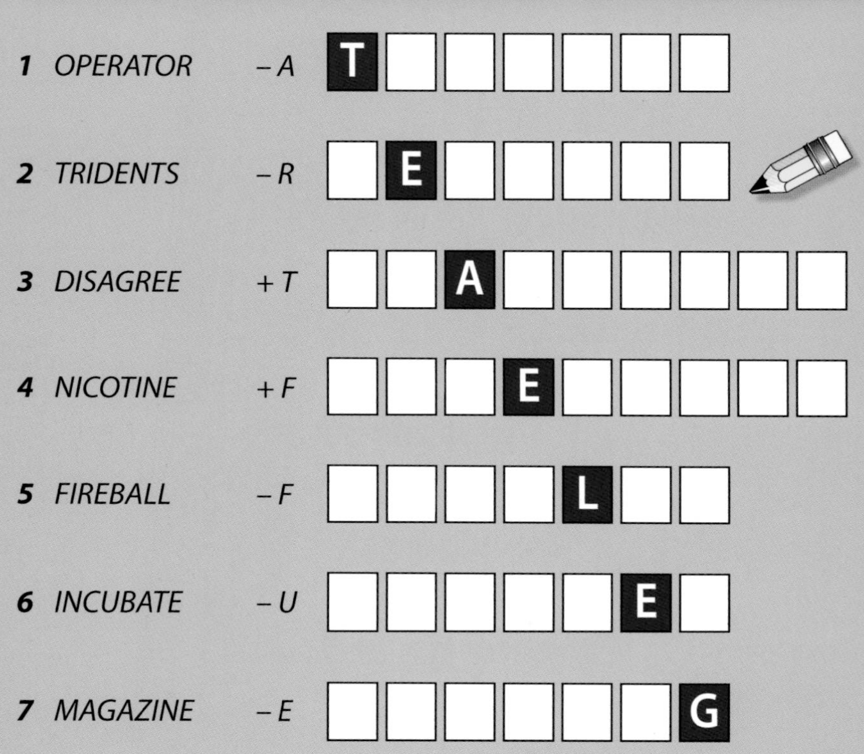

Each word on the right side contains the letters in the word at the left side, plus or minus the letter in the middle.

One letter is already in the right place.

Words & Letters

82 DOODLE PUZZLES

1 ....

2 ....

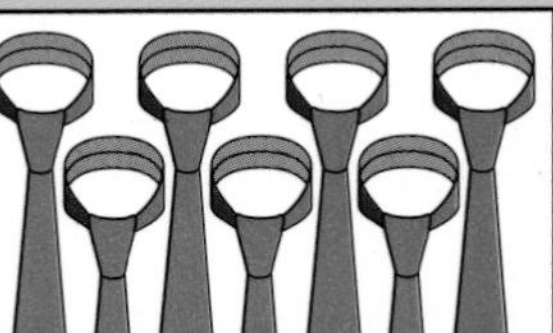

3 ....

4 ....

A doodle puzzle is a combination of images, letters, and numbers that indicate a word or concept. If you cannot solve a doodle puzzle, do not look at the answer right away. Try to solve it later or tomorrow.

Words & Letters

83 LETTER BLOCKS

1 *CRIME*

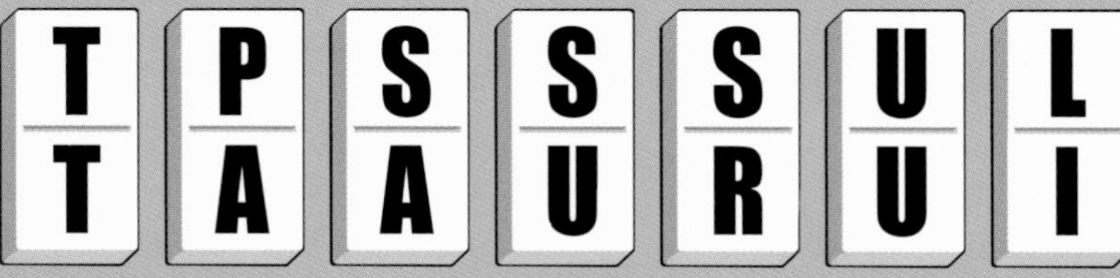

2 *CLOTHES*

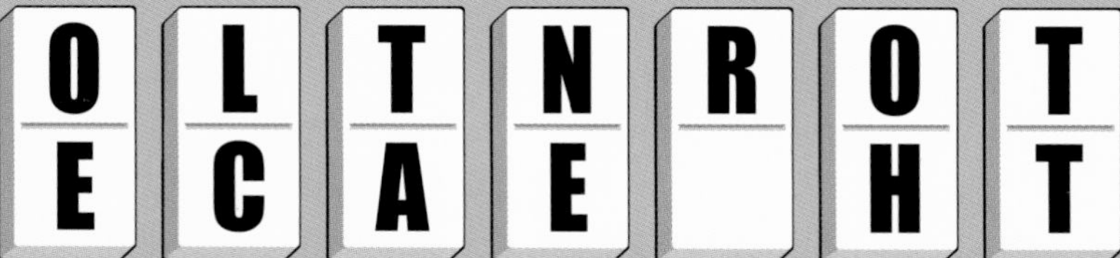

3 *SCHOOL*

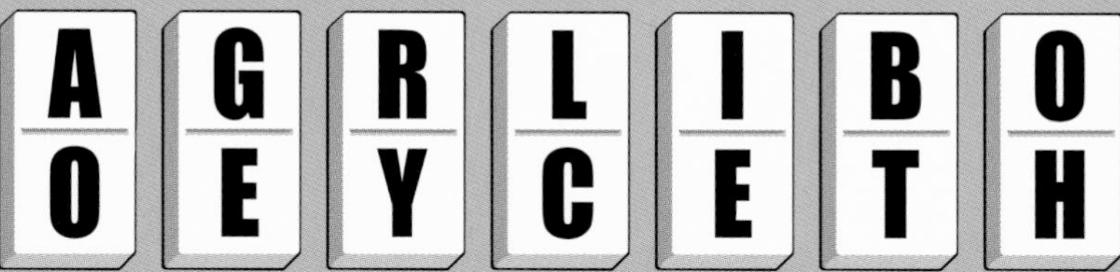

Move the letter blocks around so that words are formed ontop and below that you can associate with the subject.

The letters were reversed on two blocks.

Words & Letters

84 TEACHERS

1 *GARY HOPE*

			G					

2 *SHENG LI*

3 *CHRIS TYE*

			M					

4 *ROSY HIT*

5 *MAE SCHMIT*

		T				A				

6 *NICO COS*

E					M			

Form the course that the teacher teaches with the letters of his/her name.

One or more extra letters are already in the right place.

Short-term Memory Games

Look intently at this knot for sixty seconds, then make this knot without looking back.

Read more on page 288

4
4
5

Short-term Memory Games

It's all in the cards!

Playing Pairs, we immediately experience the limited capacity of our short-term memory. By short-term memory, we mean a period of a few seconds to a few minutes.

This work memory is needed to steer our day in the right direction. It reminds us that we just opened the faucet or just looked to the right and have stopped for a pedestrian. To put it briefly, short-term memory is important to hold on to the incredible quantity of information that we receive during the day.

We often record that information unconsciously. The information will only switch to long-term memory if the information affects you and emotions are involved.

Making associations, inventing mental notes, summoning pure concentration . . . it is essential for remembering certain information for a short time and it will certainly help you solve short-term memory games.

Short-term Memory Games

85 TEST YOUR WORD SPAN 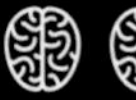

Carefully study list one, which has fifteen words. In one minute's time write down all the words you can remember.

Repeat this ten minutes later without looking at the list of words. Compare the results. If you remembered less than seven words, don't worry that your short-term memory is affected. Do not be shocked if the second attempt resulted in even fewer words. The result of this test strongly depends on your interests and ability to concentrate.

There are three lists for performing this test. You will probably obtain the best score with list two because it contains lots of emotionally charged words. With list three, you will have to assist your short-term memory.

Short-term Memory Games

85 CONTINUATION

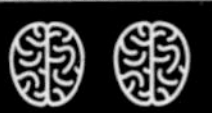

LIST 1	**LIST 2**	**LIST 3**
window	money	electricity
pencil	coffee	asphalt
dog	fire	specialist
apple pie	gold	magician
house	red	scissors
jersey	Mickey Mouse	wine
kettle	illness	aircraft
closet	mama	mat
pillow	car	dictionary
soup	sex	particle
telephone	pool	pond
train	Obama	tulip
tree	Iphone	paper
computer	music	mosaic
radio	dream	analysis

Short-term Memory Games

86 IN ALPHABETICAL ORDER

A 1

B 2

C 3

D 4

. . .

Z 26

Say the entire alphabet out loud. After each letter, say its position in the alphabet without using a pen and paper.

Restart as soon as you make a mistake.

Short-term Memory Games

87 SHAPES, COLORS, SIZES, . . .

Look intently at these colored shapes for sixty seconds, then complete the assignment on the back of this page.

Short-term Memory Games

87 CONTINUATION

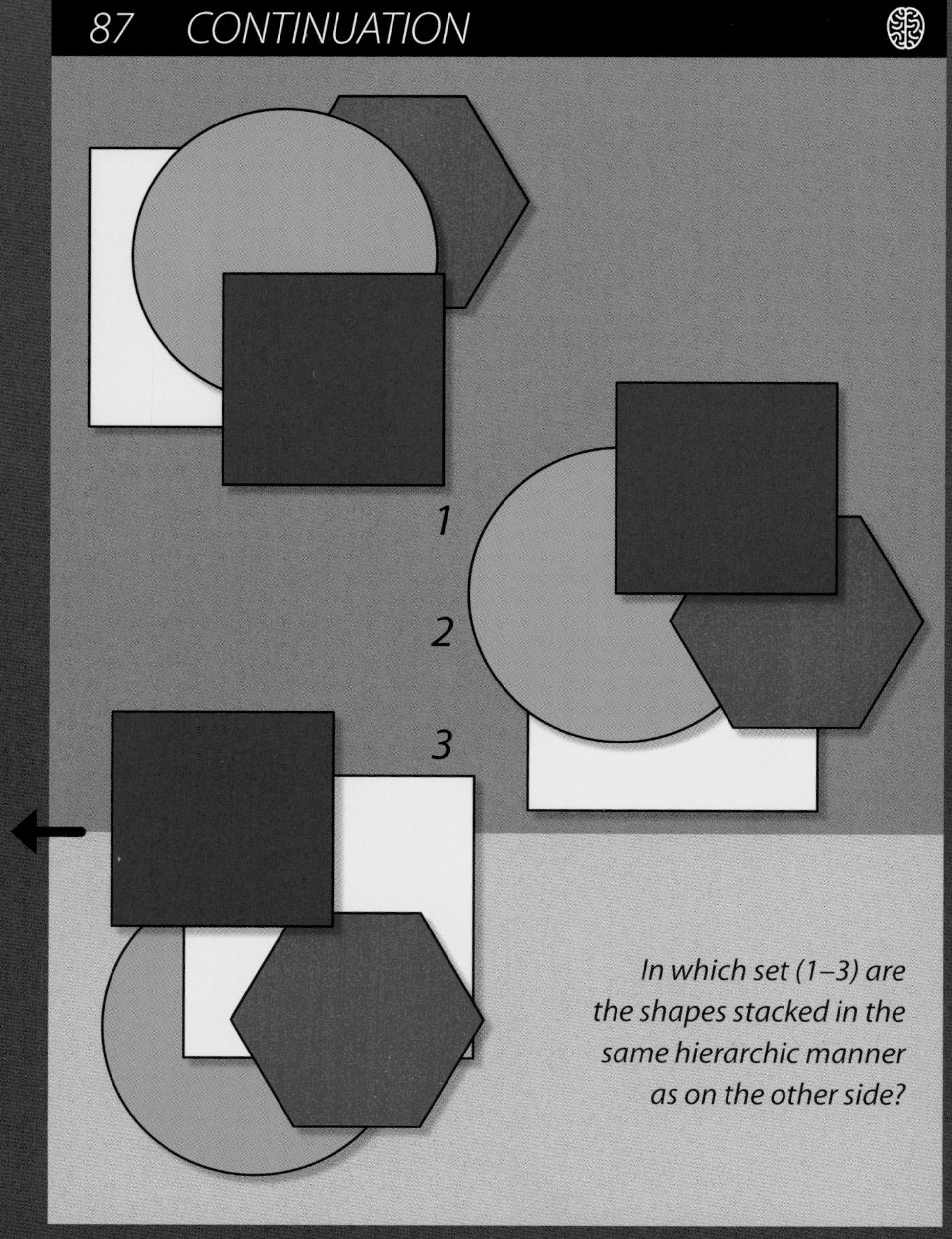

In which set (1–3) are the shapes stacked in the same hierarchic manner as on the other side?

Short-term Memory Games

88 ILLUSION

Look intently at these three impossible figures for two minutes, then complete the assignment on the back of this page.

→

Short-term Memory Games

88 CONTINUATION

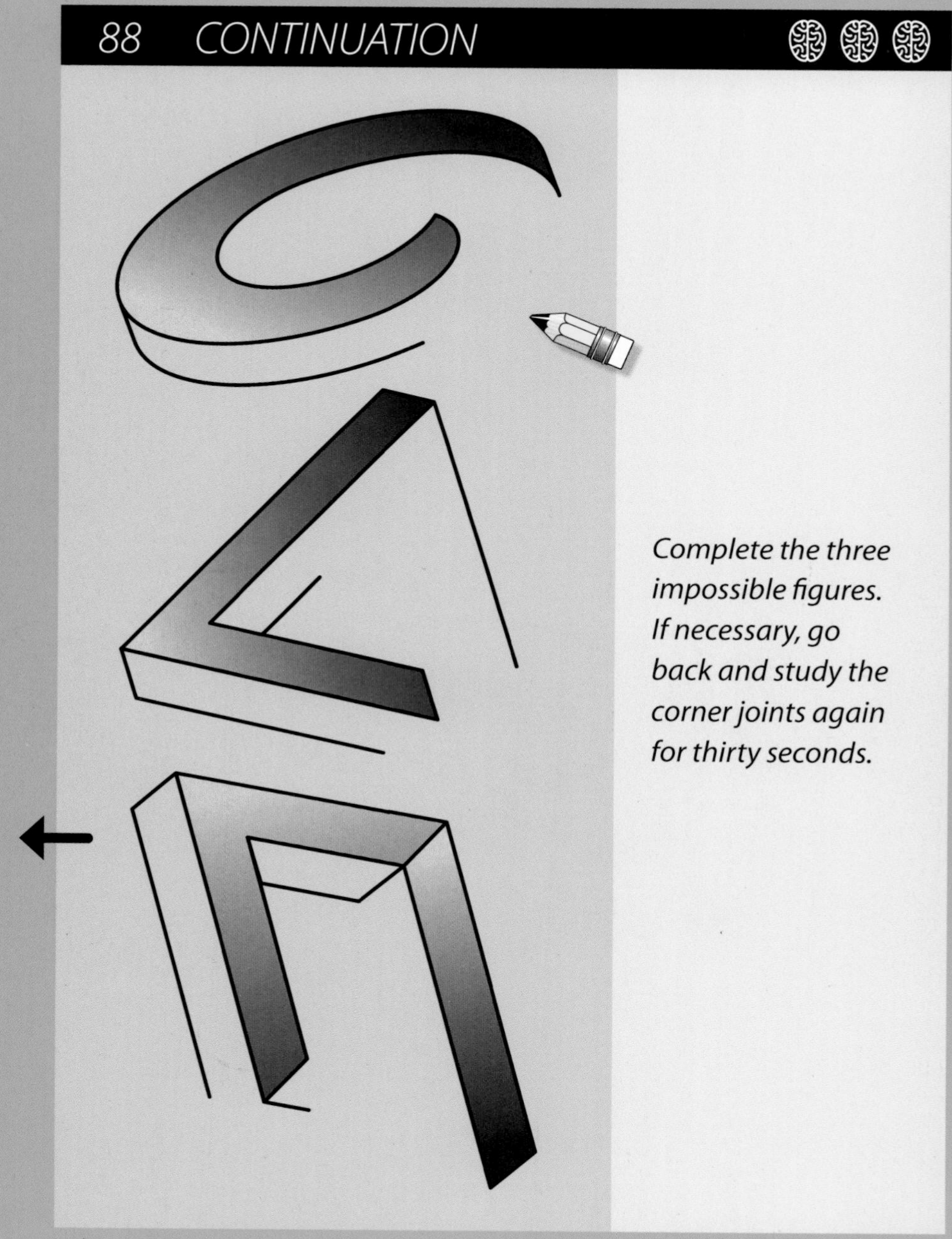

Complete the three impossible figures. If necessary, go back and study the corner joints again for thirty seconds.

Short-term Memory Games

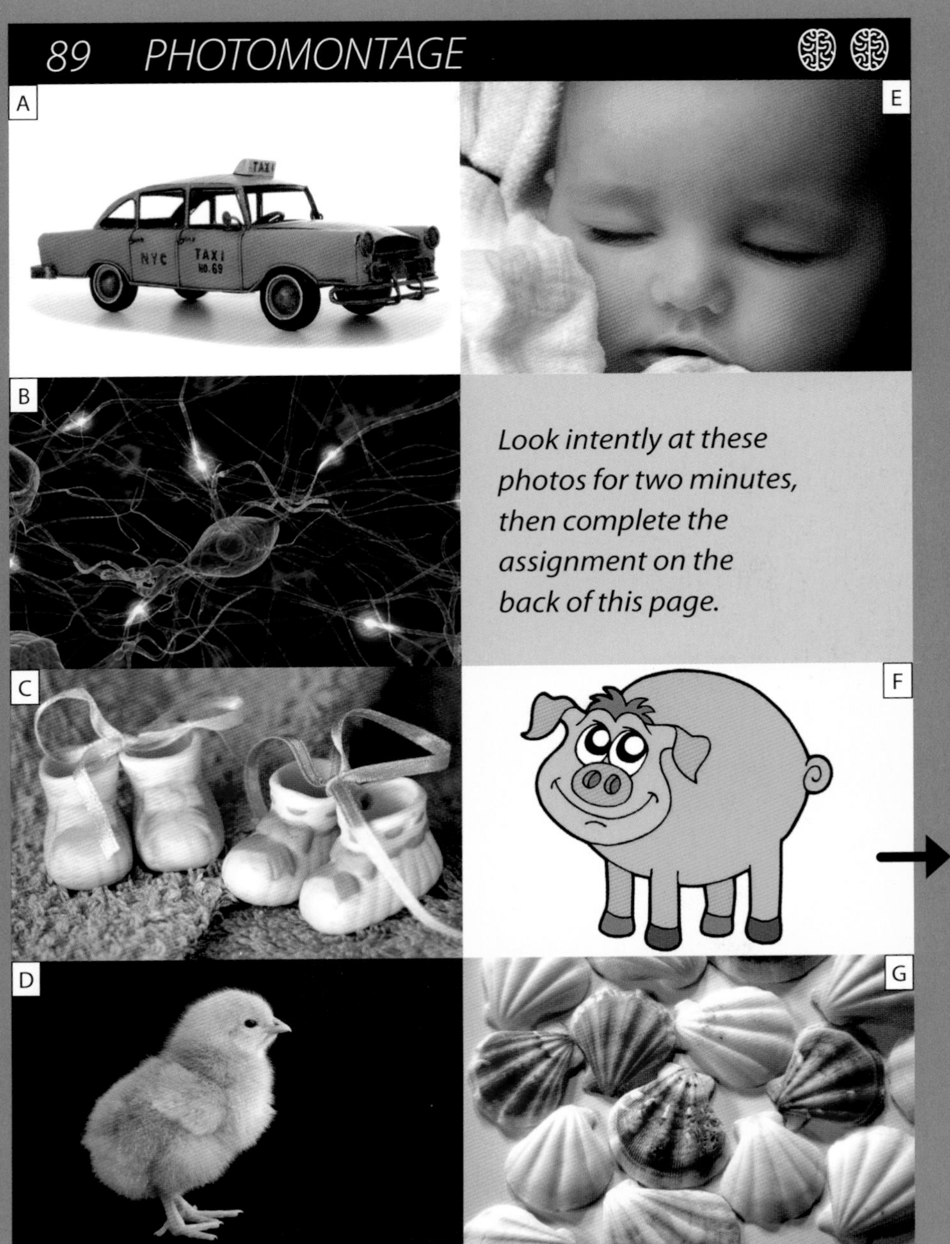

89 PHOTOMONTAGE
A
TAXI
NYC
TAXI
NO.69
E
B
Look intently at these photos for two minutes, then complete the assignment on the back of this page.
C
F
D
G

Short-term Memory Games

89 CONTINUATION

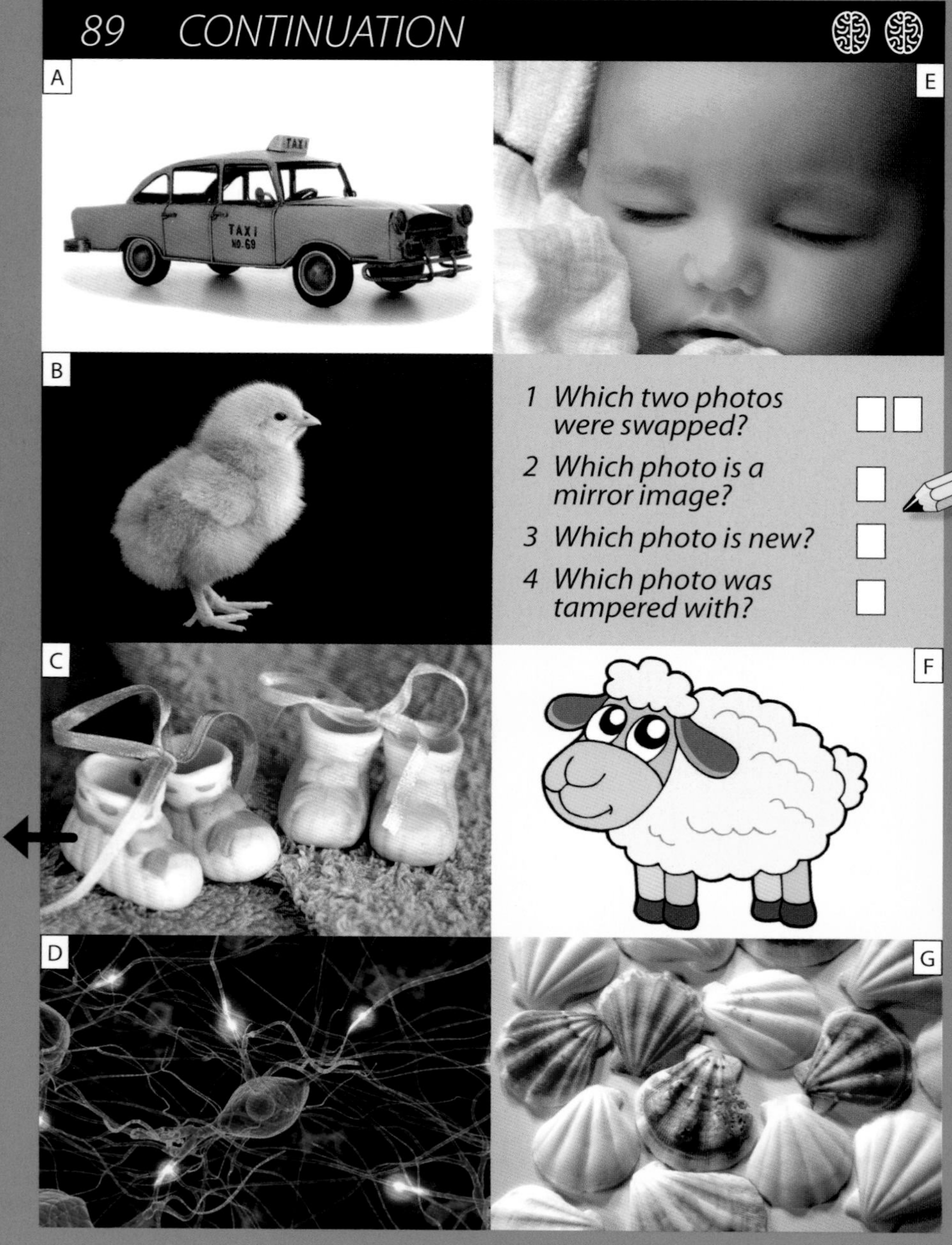

1 Which two photos were swapped?

2 Which photo is a mirror image?

3 Which photo is new?

4 Which photo was tampered with?

Short-term Memory Games

90 FIND THE DIFFERENCES

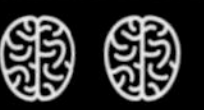

Look intently at all elements for sixty seconds, then complete the assignment on the back of this page.

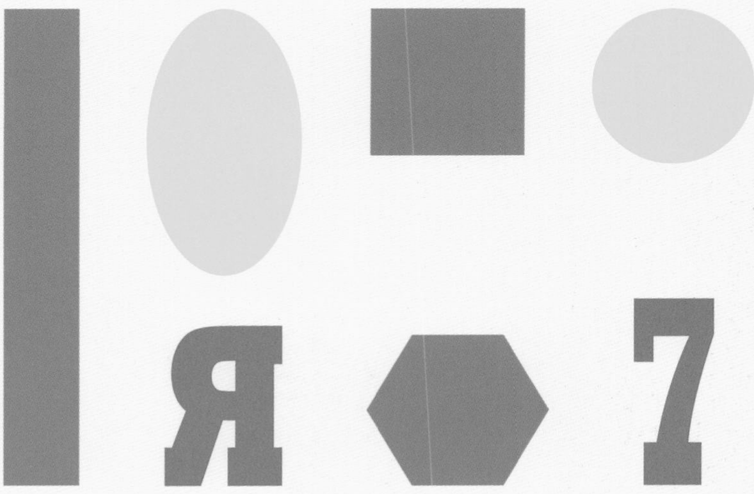

Look intently at these elements for sixty seconds, then complete the assignment on the back of this page.

Short-term Memory Games

90 CONTINUATION

Look intently at all seven elements for sixty seconds, then complete the assignment on the back of this page.

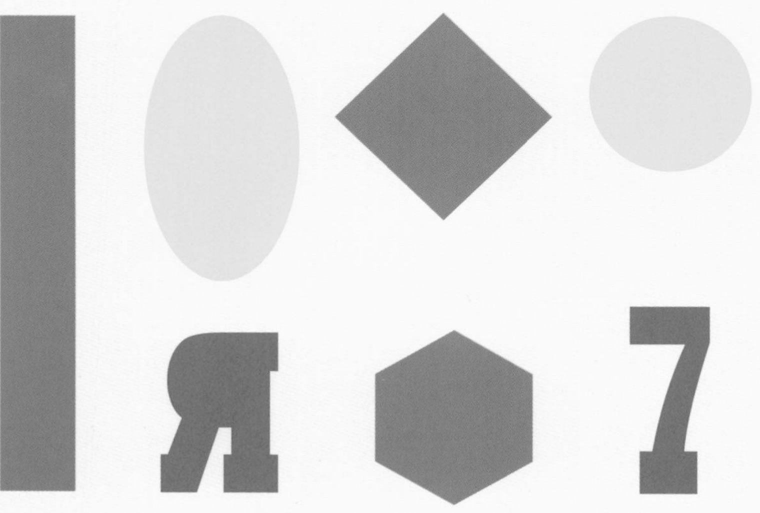

Look for the five errors.

Short-term Memory Games

91 BIG DISCOUNTS

–8%

–20%

2 for 1

–5%

–33%

Look intently at this tasty food and super discounts for two minutes, then complete the assignment on the back of this page.

Short-term Memory Games

91 CONTINUATION

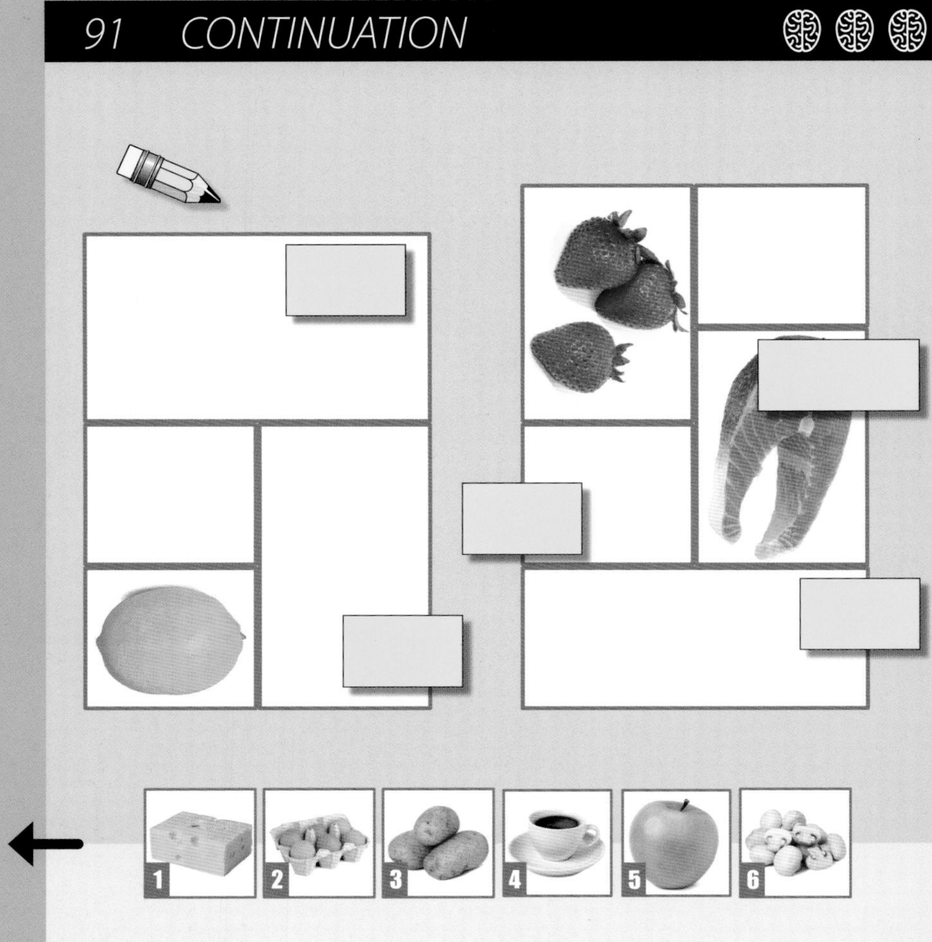

Now you are looking from a different direction. Enter the missing foods (1–6) and list all discounts next to the correct products without rotating the page one hundred eighty degrees.

Short-term Memory Games

92 SPECIAL TRIVIA

1. *What is Obelix's occupation?*
2. *Where is Lake Titicaca?*
3. *Where does the polar bear live?*
4. *What is the name of the river where a Roman general had to disband his army before continuing on to Rome?*
5. *What is the name of the science of wine in agriculture?*
6. *What is the name of an abbreviation that is not read as one word but as separate letters?*
7. *If you list the presidents of the United States, which number is Obama?*
8. *Where are 2 by 3 dots grouped on grids so that a total of 63 signs are possible?*
9. *Which hero can crawl across the ceiling?*
10. *What is the scientific name of the pole star?*

→

Try to remember these multiple-choice questions so that you can answer them correctly on the back without returning to this page.

Note: the multiple-choice answers on the back are in random order.

Short-term Memory Games

92 CONTINUATION

○ Po	○ Nordicum	○ the 42nd
○ viticulture	○ South Pole	○ synonym
○ carpenter	○ initials	○ in Italy
○ Morse alphabet	○ vacuum cleaner	○ stenography
○ the 44th	○ Pole sun	○ Tiber
○ in Chili	○ farmer	○ in Peru
○ independence	○ oenology	○ Spiderman
○ Superman	○ acronym	○ Batman
○ North Pole	○ vinology	○ Polaris
○ Rubicon	○ menhir sculptor	○ braille

Indicate the correct answers to the ten questions.

Short-term Memory Games

93 COPY PASTE

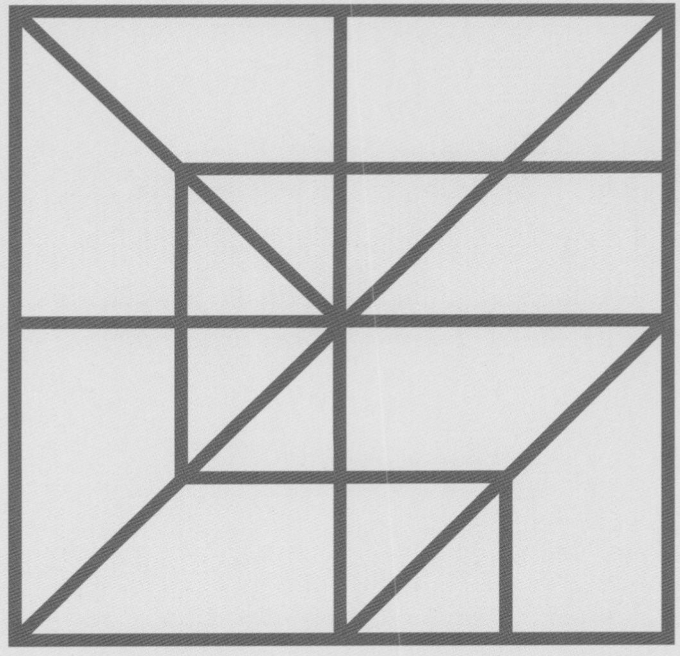

Look intently at the composition above for sixty seconds, then complete the assignment on the back of this page.

Short-term Memory Games

93 CONTINUATION

Redraw the figure on the previous page along the dotted lines.

Short-term Memory Games

94 NATO's SPELLING ALPHABET 

Alpha	**Juliet**	**Sierra**
Bravo	**Kilo**	**Tango**
Charlie	**Lima**	**Uniform**
Delta	**Mike**	**Victor**
Echo	**November**	**Whiskey**
Foxtrot	**Oscar**	**X-ray**
Golf	**Papa**	**Yankee**
Hotel	**Quebec**	**Zulu**
India	**Romeo**	

→

NATO's spelling alphabet is used for spelling messages with as few errors as possible, even if the phone connection is very bad. It is constructed so that words from the alphabet cannot be confused, even if a Frenchman needs to send a message to a Greek.

Try to remember them all, like a good soldier, then read the assignment on the back.

Short-term Memory Games

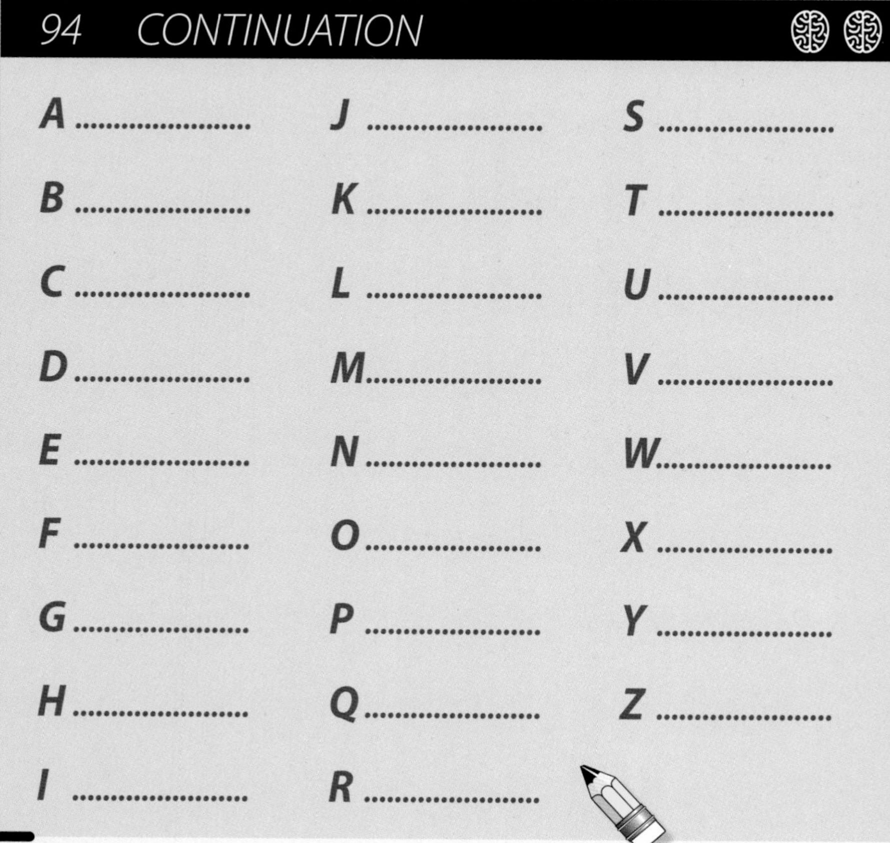

94 CONTINUATION
A ....
B ....
C ....
D ....
E ....
F ....
G ....
H ....
I ....
J ....
K ....
L ....
M ....
N ....
O ....
P ....
Q ....
R ....
S ....
T ....
U ....
V ....
W ....
X ....
Y ....
Z ....

The assignment is simple.
Next to each letter of the alphabet, place the corresponding word from NATO's spelling alphabet.

Short-term Memory Games

95 COLORING PICTURE

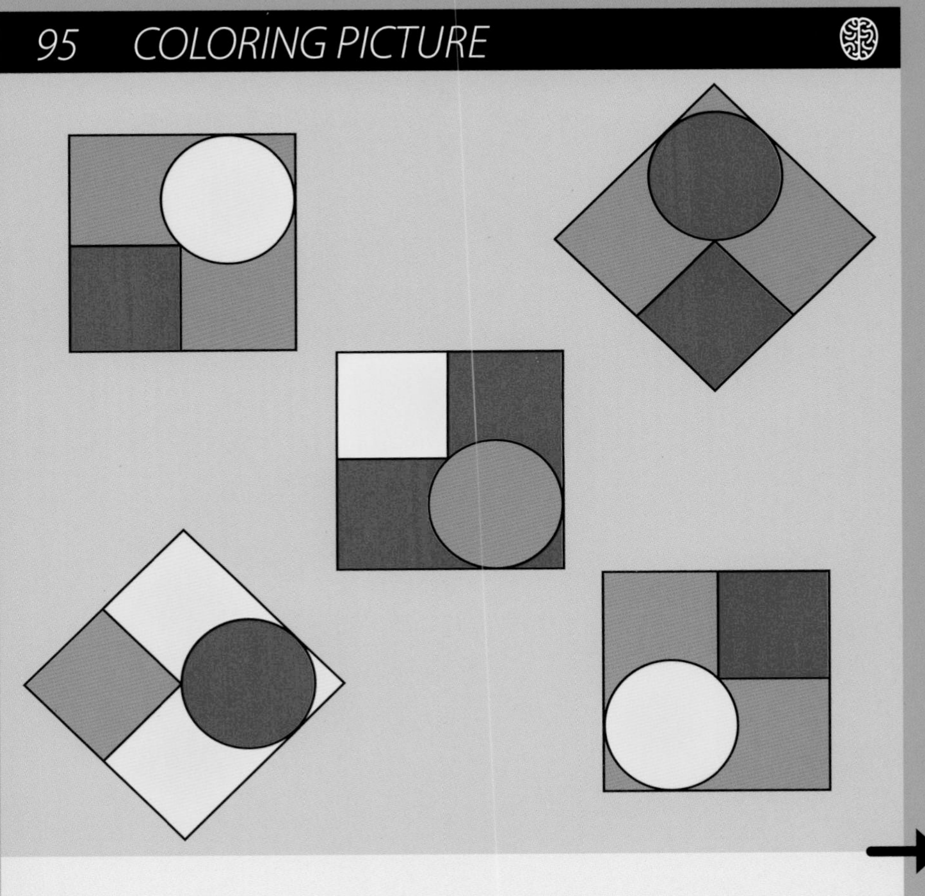

Look intently at the composition above for sixty seconds, then complete the assignment on the back of this page.

Short-term Memory Games

95 CONTINUATION

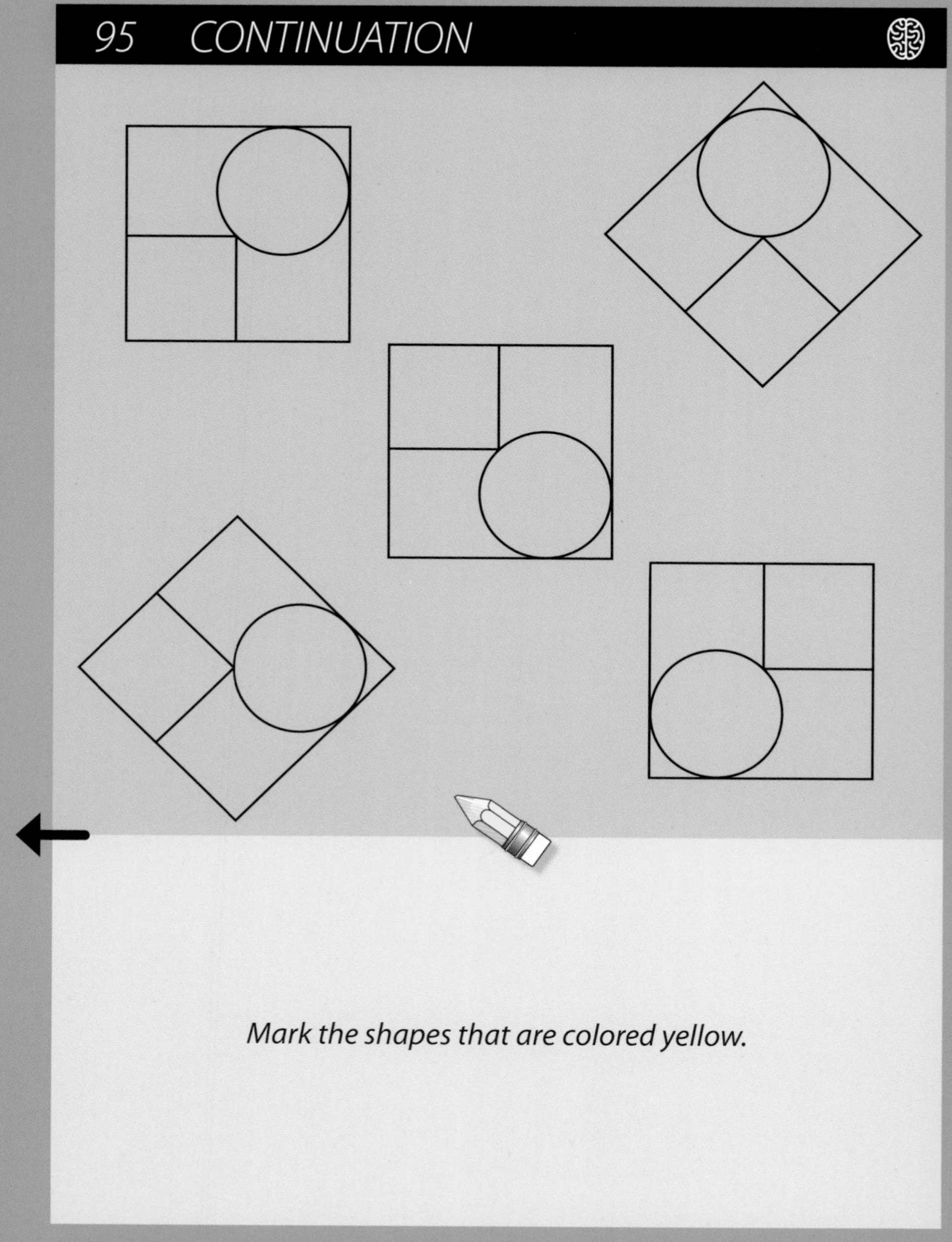

Mark the shapes that are colored yellow.

Short-term Memory Games

96 HEALTHY FISH

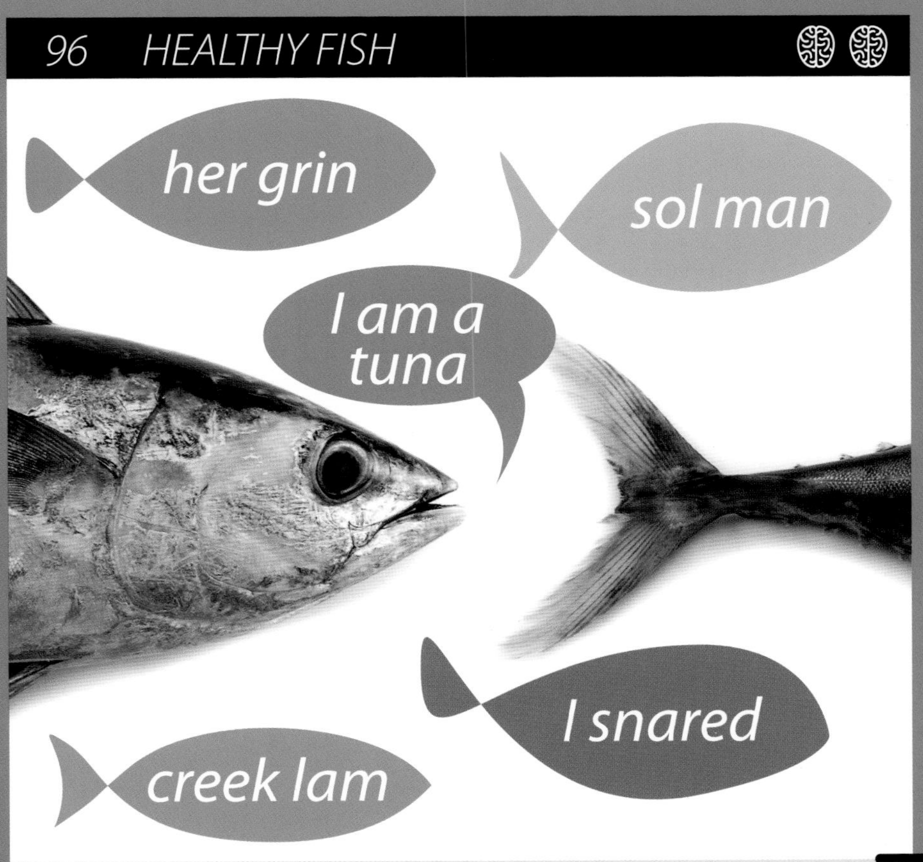

Look intently at these weird fish names for sixty seconds, then complete the assignment on the back of this page.

Short-term Memory Games

96 CONTINUATION

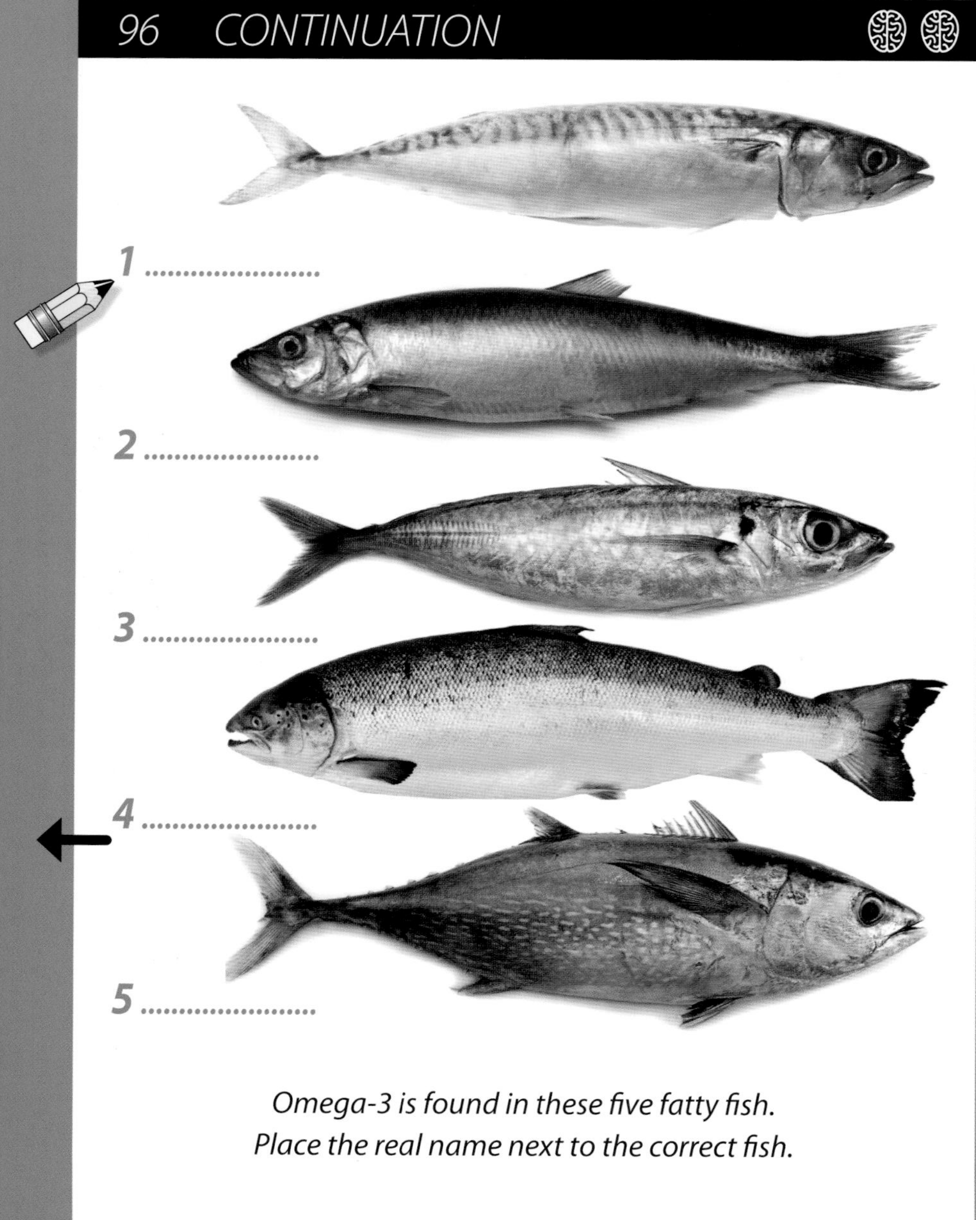

1 ....

2 ....

3 ....

4 ....

5 ....

Omega-3 is found in these five fatty fish.
Place the real name next to the correct fish.

Short-term Memory Games

97 ROLL THE DICE

Look intently at these dice for sixty seconds, then complete the assignment on the back of this page.

97 CONTINUATION

Color in the unnecessary spots on the changed dice.

Short-term Memory Games

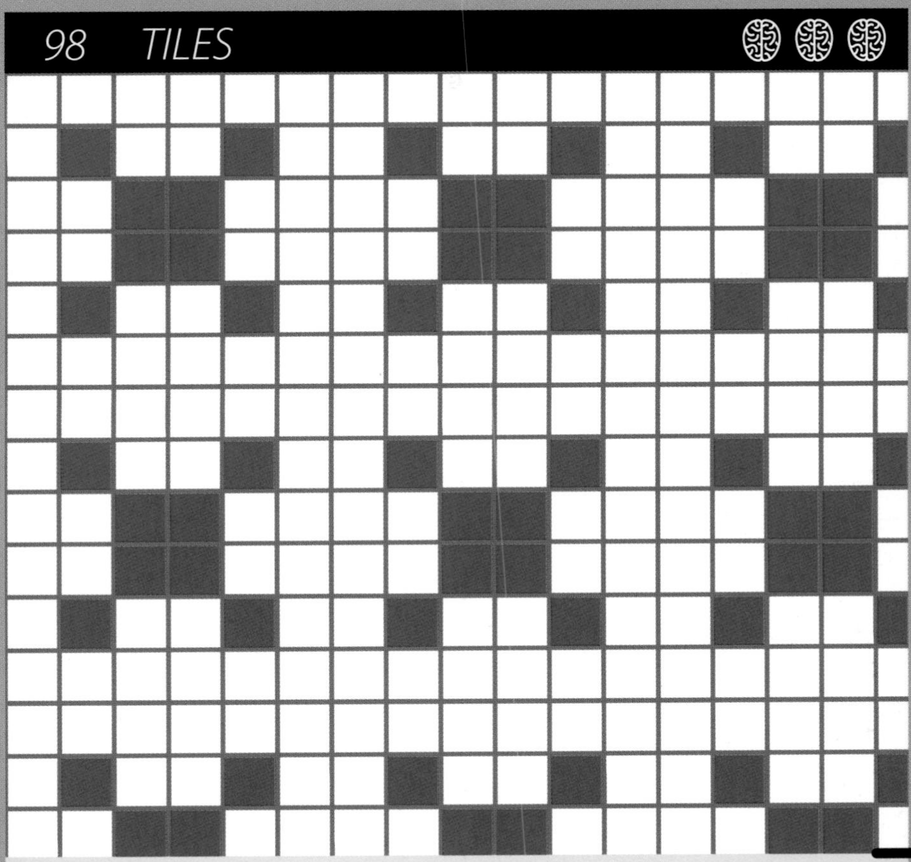

Look intently at this pattern for sixty seconds, then complete the assignment on the back of this page.

Short-term Memory Games

98 CONTINUATION

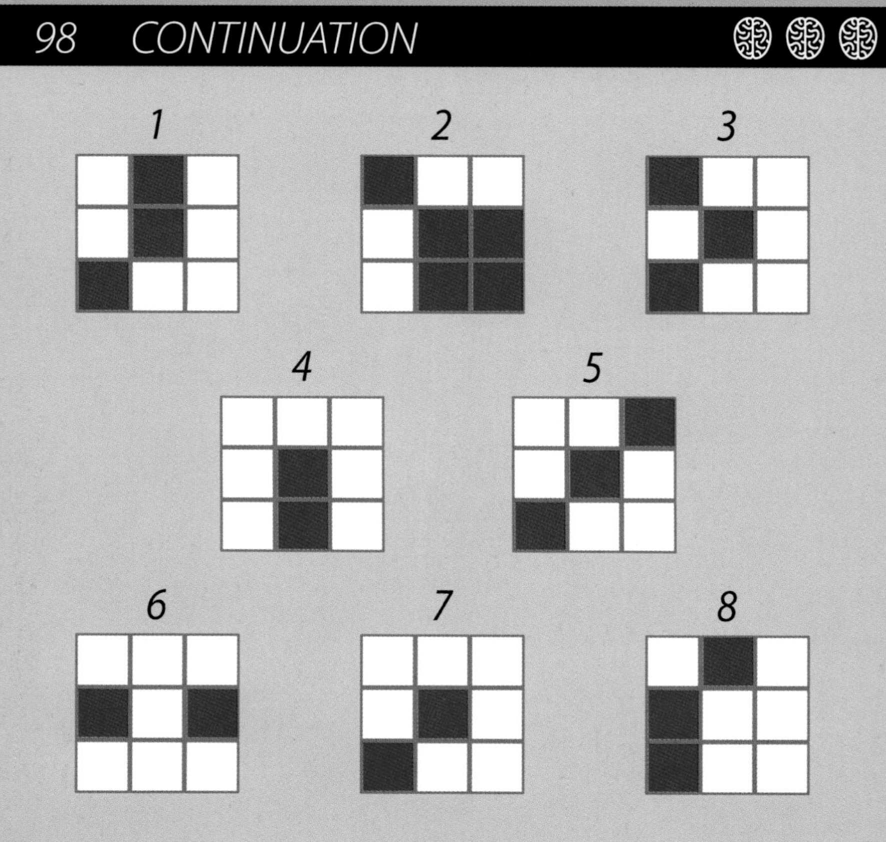

With which tile (1–8) can you lay the floor that is shown on the previous page?

Short-term Memory Games

99 SPECIAL FIVE

2 3 6

1 5 4

8 7 9

Look intently at these numbers for sixty seconds, then complete the assignment on the back of this page.

Short-term Memory Games

99 CONTINUATION

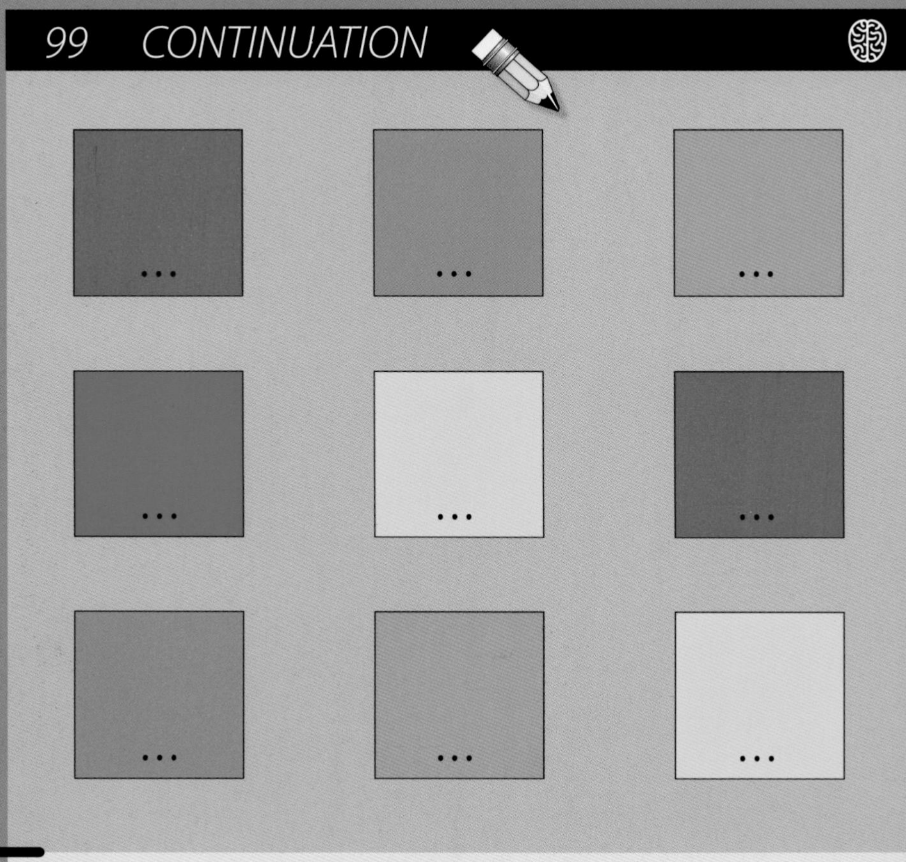

1.

What is the sum of all figures?

2.

Write the identically colored numbers in the colored squares.

Go back if necessary and study the colors between the numbers.

Short-term Memory Games

100 NOT IN ONE, TWO, THREE  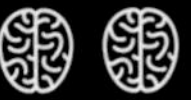

**How much is
fifty-four thousand,
three hundred twenty-one
minus twelve thousand,
three hundred forty-five?**

**How much is
one thousand, two hundred
thirty-four plus
four thousand, three hundred
twenty-one?**

Mentally calculate the two assignments above.

3-D FUN

Which letters of the alphabet look the same when turned 180 degrees.

Solution on page 288

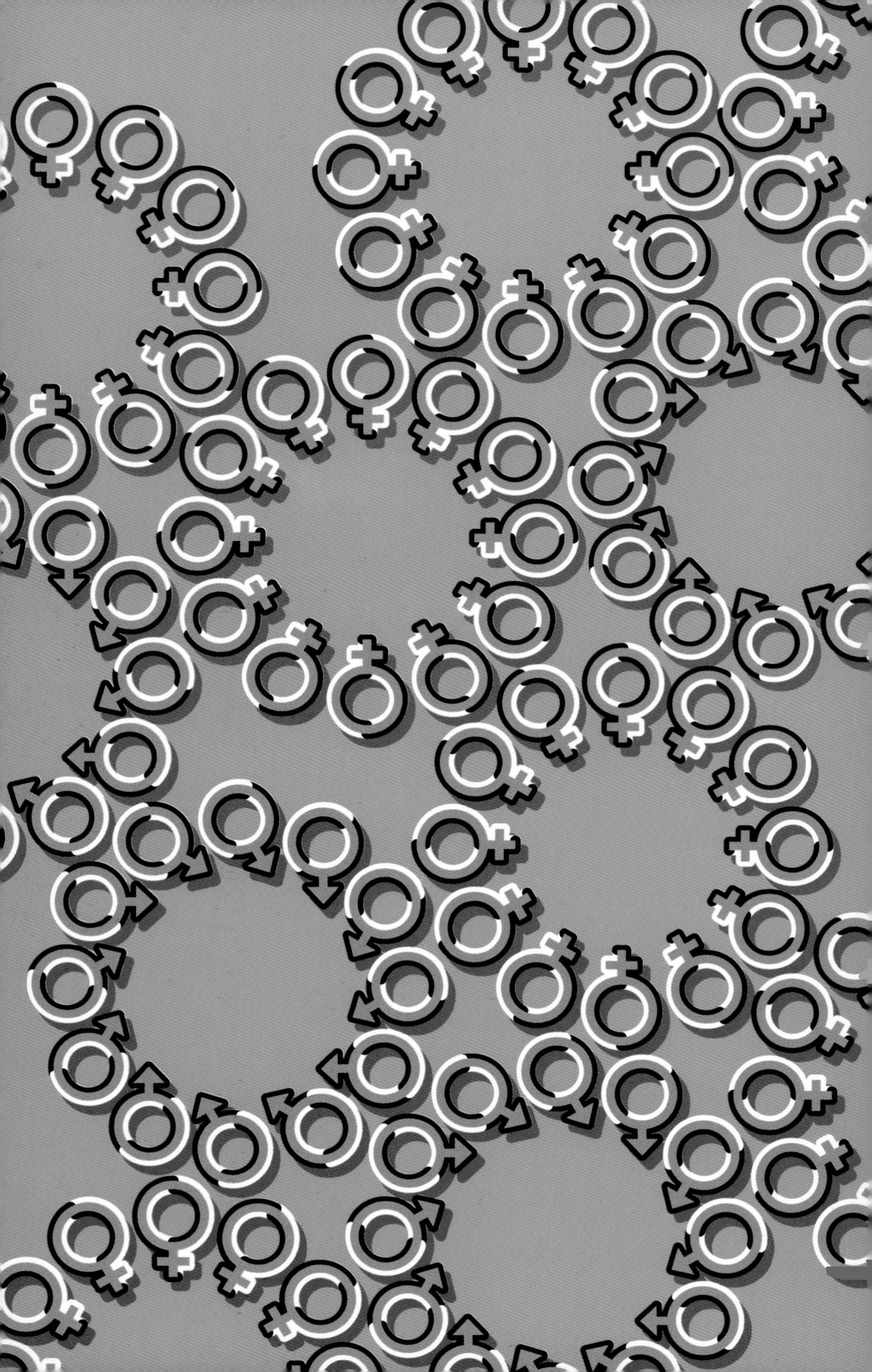

Honey, it is the next street on the right. No, it is on the left!

One difference between men and women is that men have better spatial visualization ability and can visualize objects better. You will have to figure out whether the following differences between men and women are to your advantage when solving 3-D puzzles.

Women

- make more use of the collaboration between the left and right hemispheres
- have a more well-developed visual memory
- usually have more well-developed fine motor skills
- can remember series of incoherent words better
- are better able to execute several tasks at the same time
- have a wider field of vision than men
- can recognize faces better
- can better express their emotions
- are more susceptible to depressions

Men

- can better tackle complex technical problems
- are more action-oriented in problem situations
- can read maps better due to a a more well-developed special visualization ability
- have a field of vision that reaches farther
- can better concentrate on one target
- are better at saving classified information
- are less susceptible to depressions

3-D Fun

101

Which cube (1–6) fits in the empty corner?

3-D Fun

102

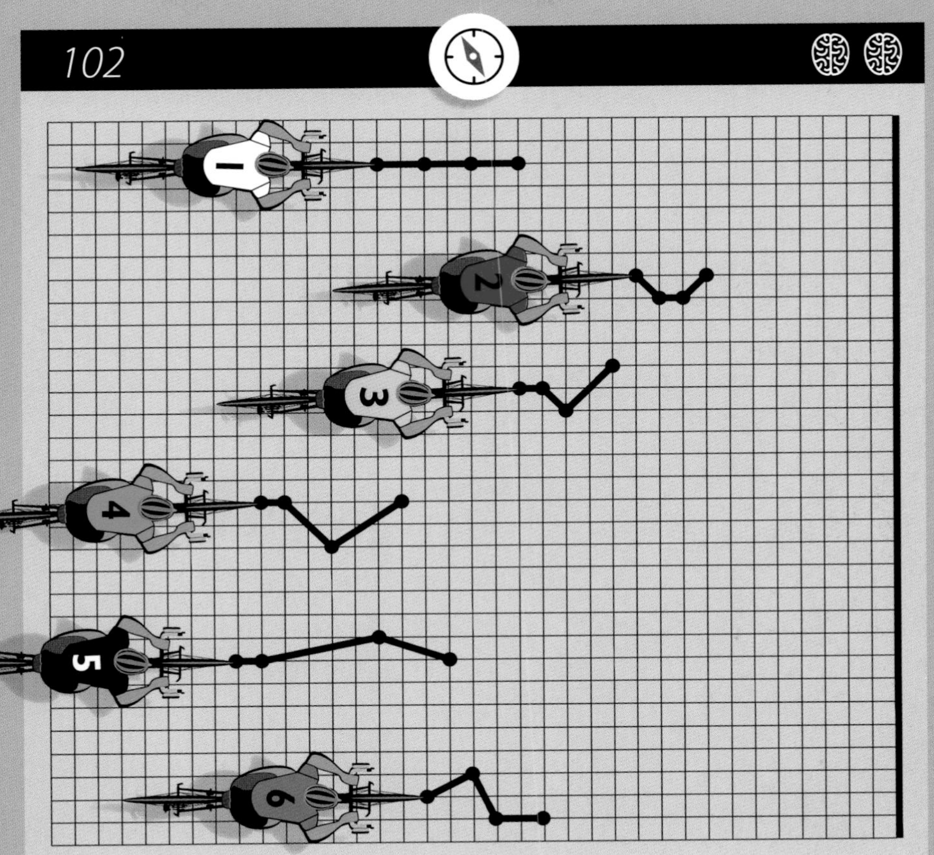

The grid above shows the distance every rider covers in three seconds. Each dot represents his progress per second.

Which rider (1–6) will reach the finish line first?

3-D Fun

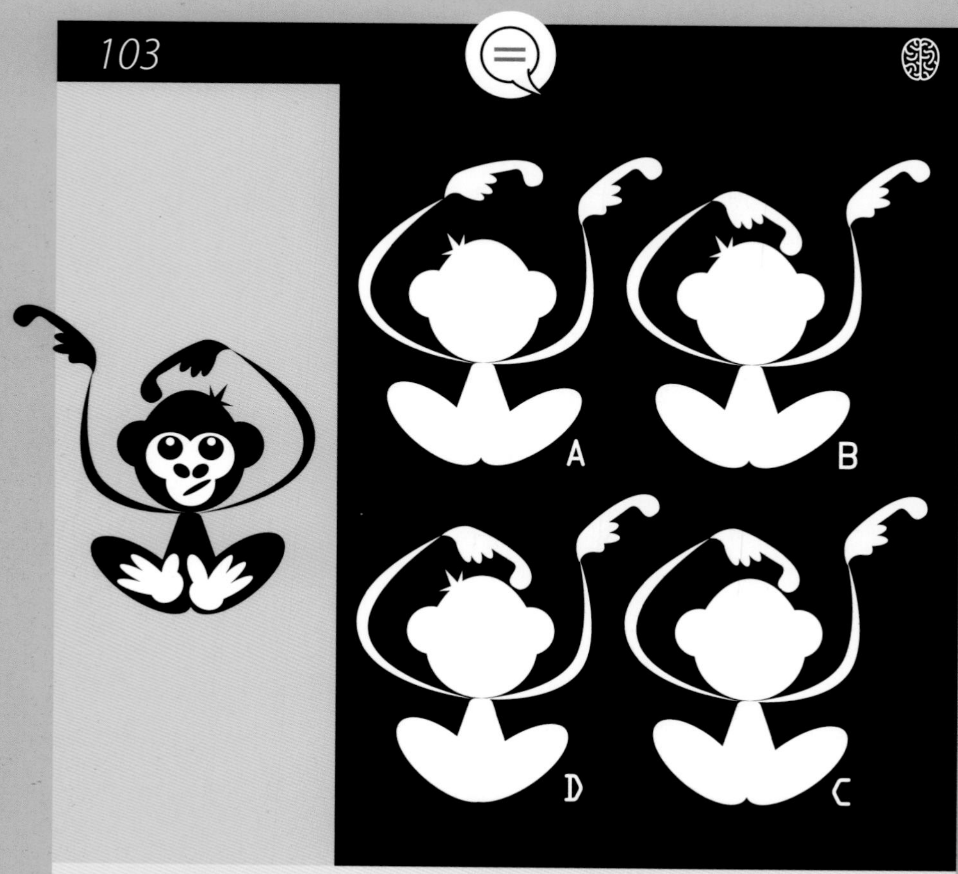

Which back (A–D) corresponds with the image of the monkey?

3-D Fun

104

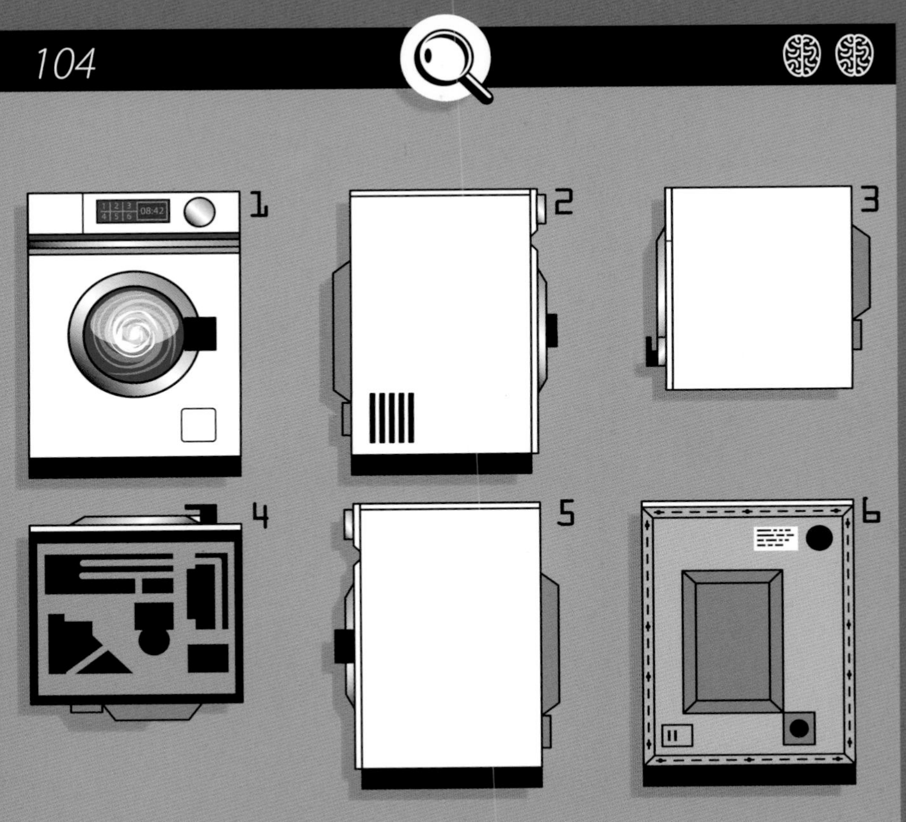

Above is a washing machine from six different angles.

Which angle (1–6) is shown incorrectly?

3-D Fun

105

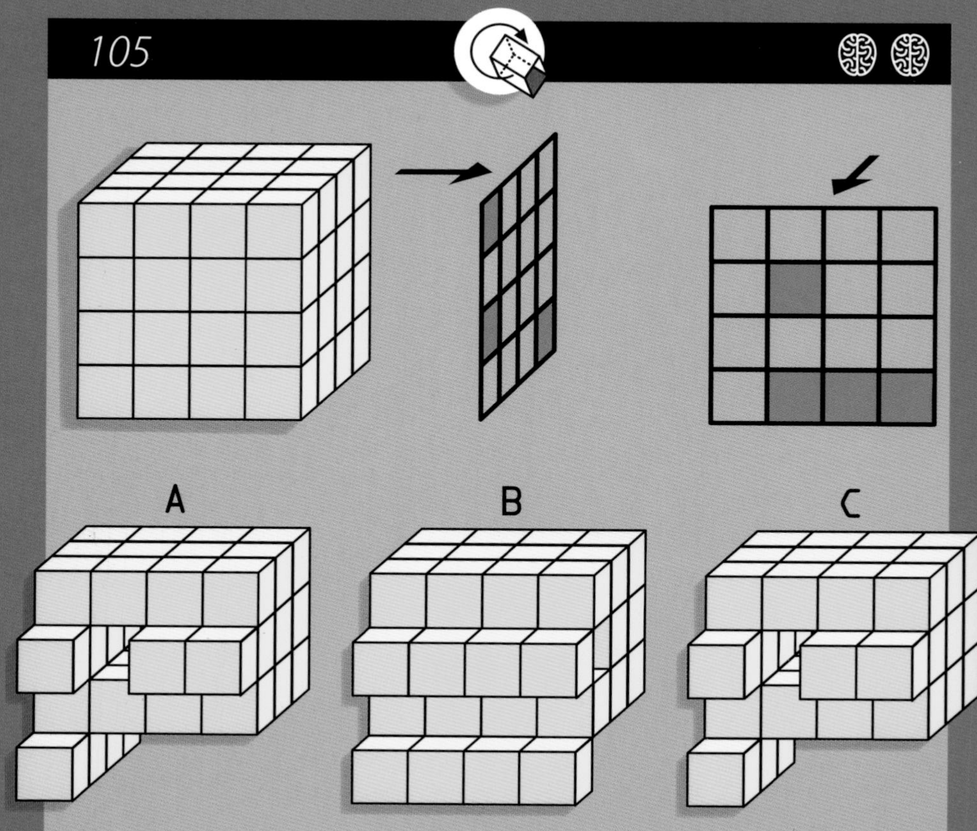

A piece of cheese successively passes through two molds. The piece of cheese will only go where there is no gray face. The cheese will move from left to right through the first mold and from back to front through the second mold.

Which piece of cheese (A–C) will we be left with?

3-D Fun

106

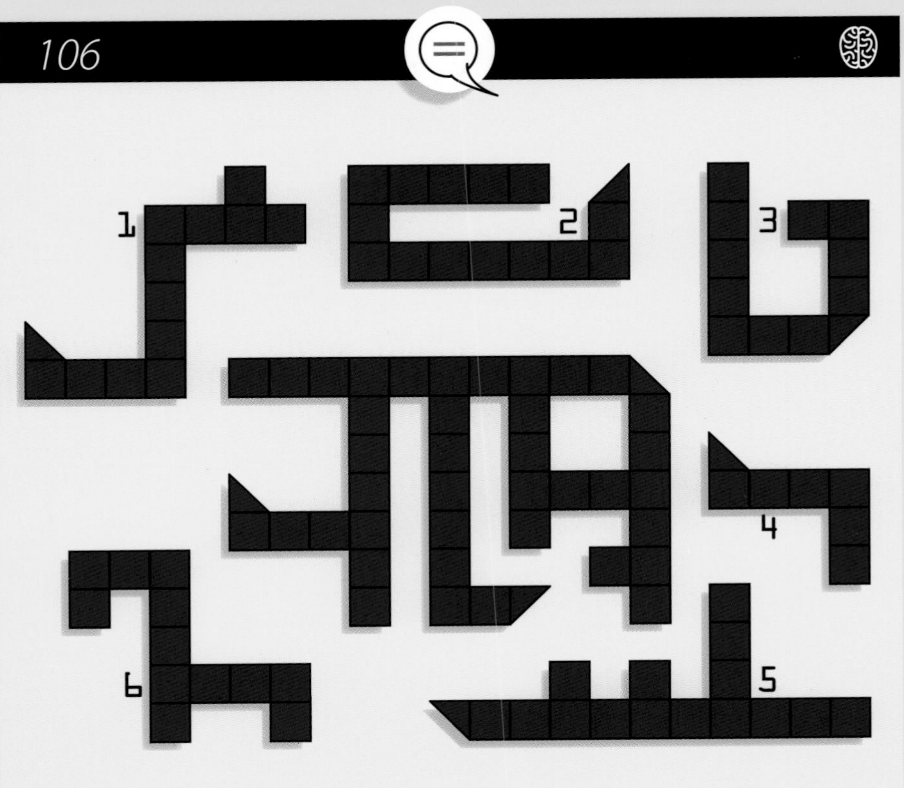

Which three elements (1–6) fit on top of the base structure in the middle? The shapes can be rotated but not mirrored.

3-D Fun

107

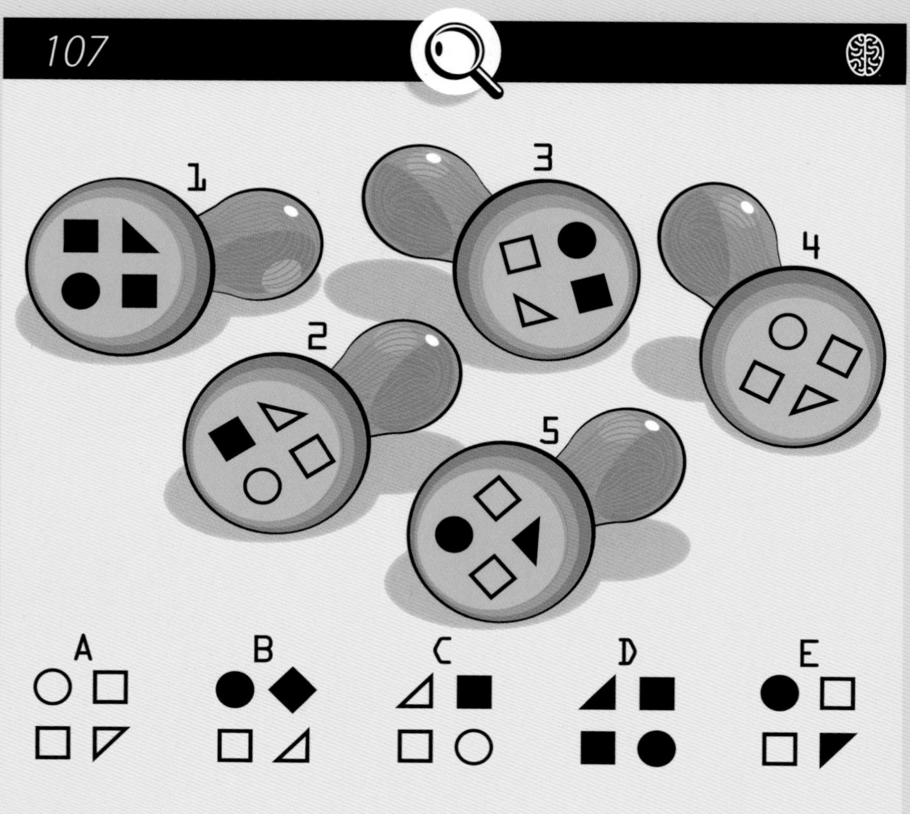

Which stamp (A–E) does not originate from any of the rubber stamps above (1–5)?

3-D Fun

108

Which surface (1–13) is colored incorrectly?

3-D Fun

109

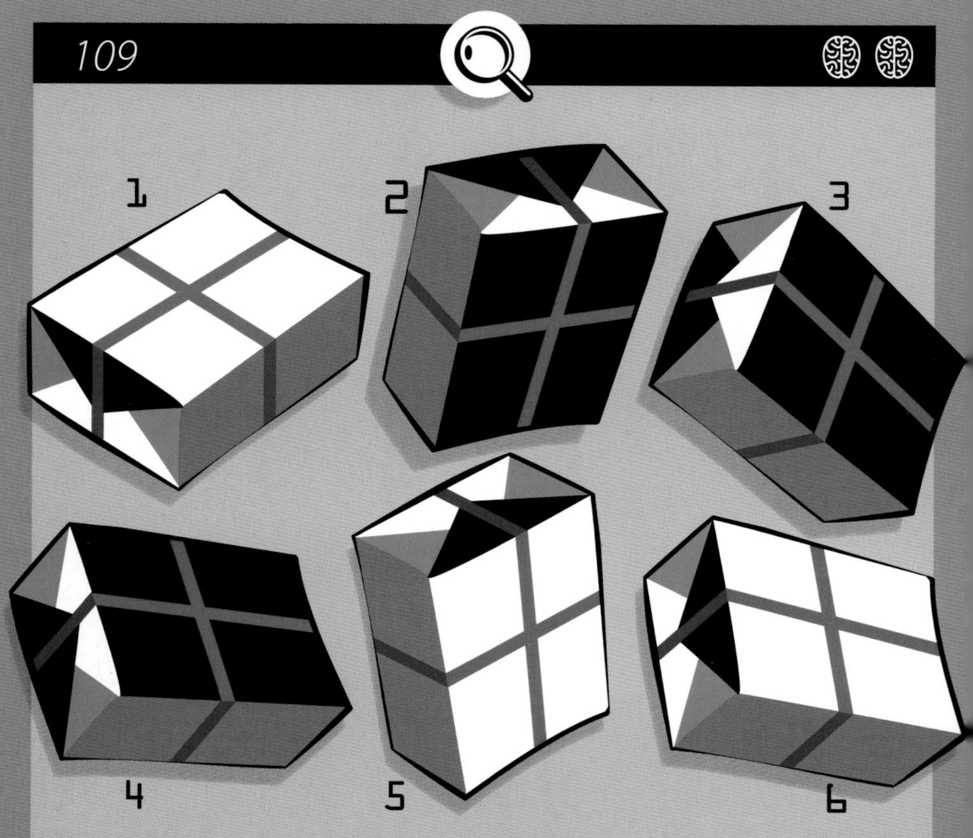

Which gift (1–6) does not belong?

3-D Fun

110

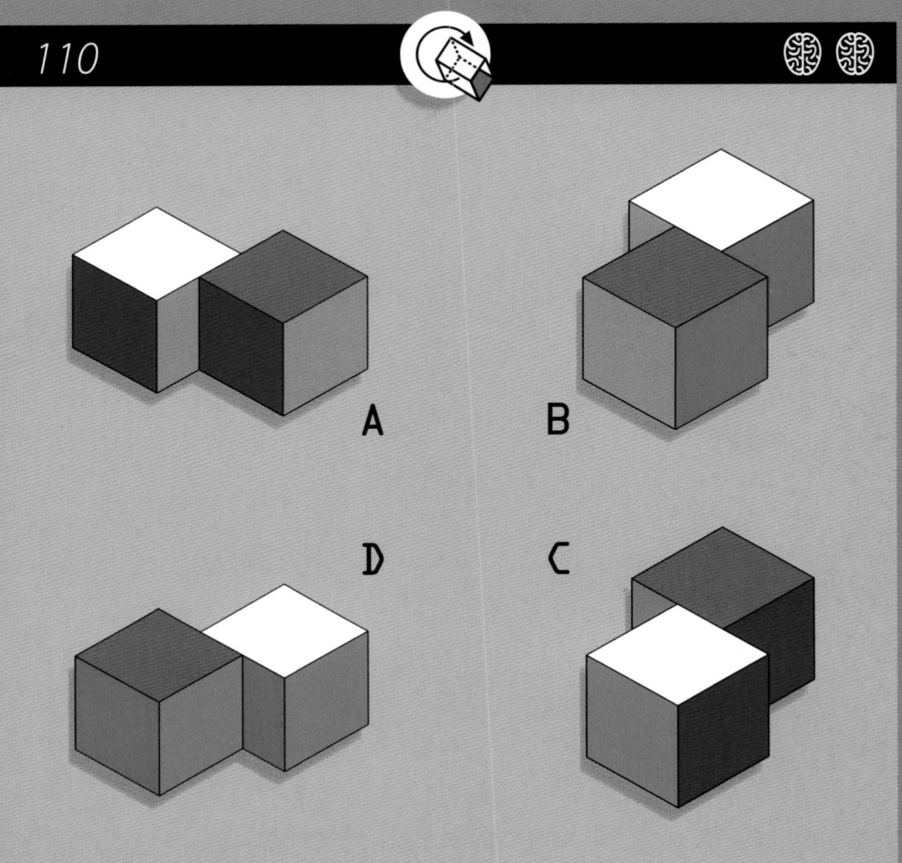

Which profile (A–D) of the same construction with cubes is wrong?

3-D Fun

111

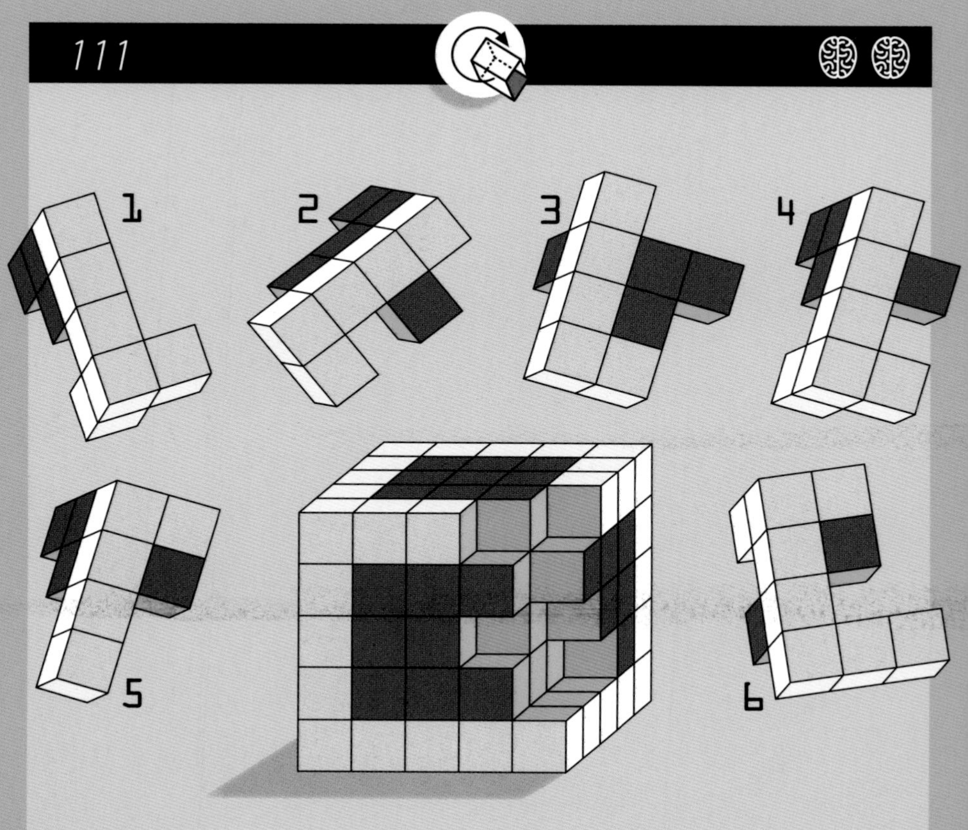

Which piece (1–6) completes the cube?

3-D Fun

112

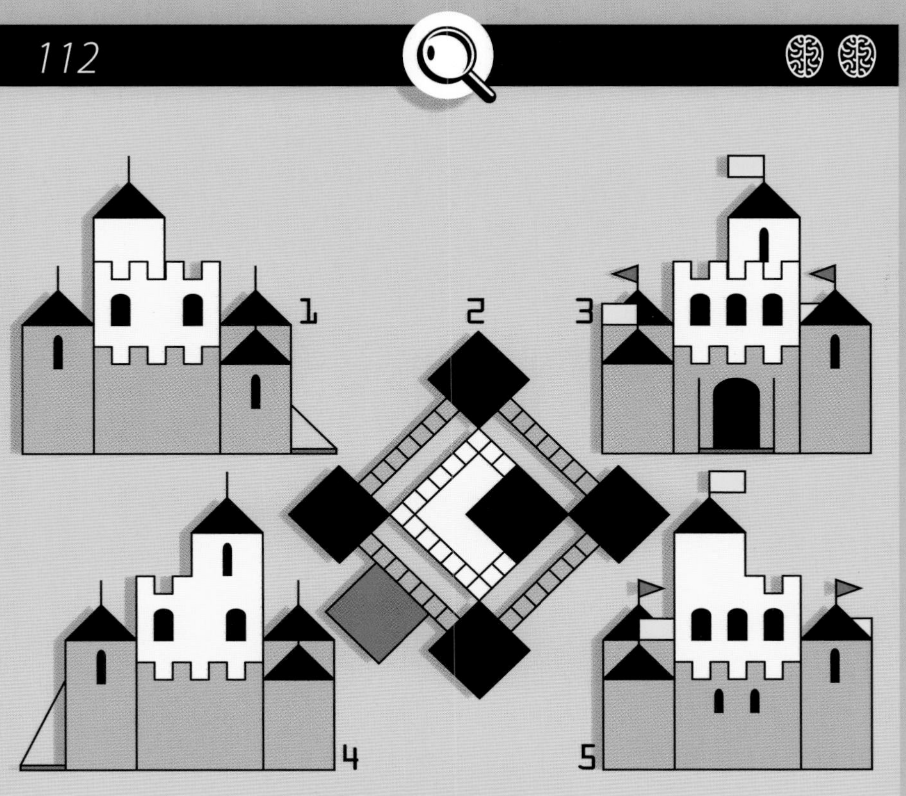

Which angle (1–5) of this castle is wrong?

3-D Fun

113

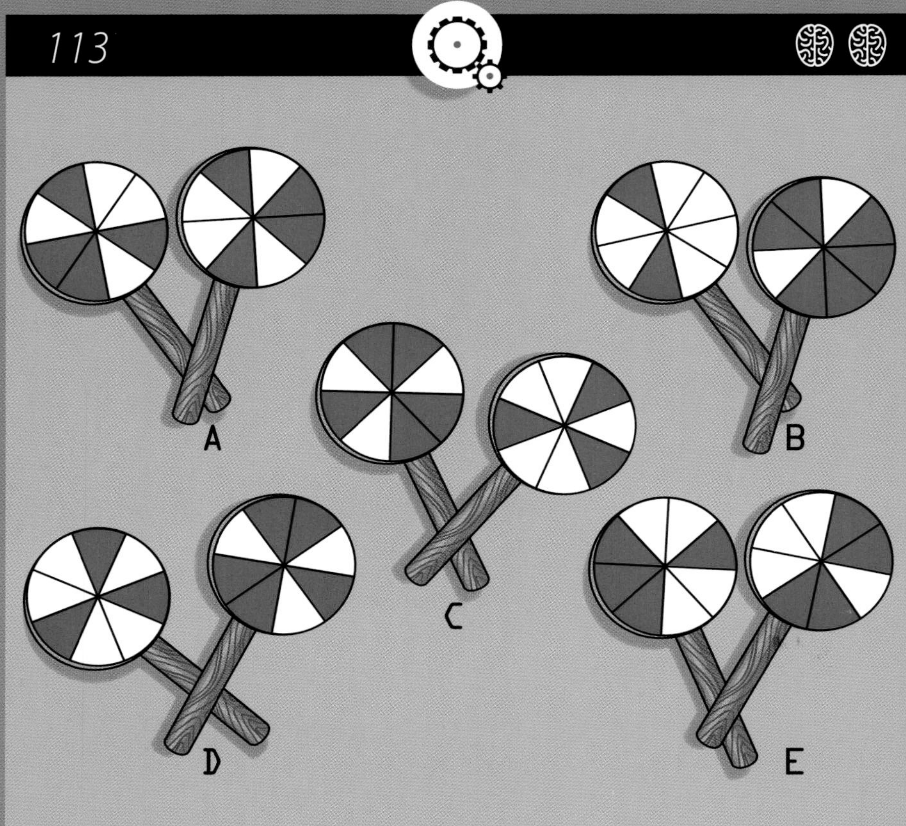

Which pair of signs (A–E) does not belong?

3-D Fun

114

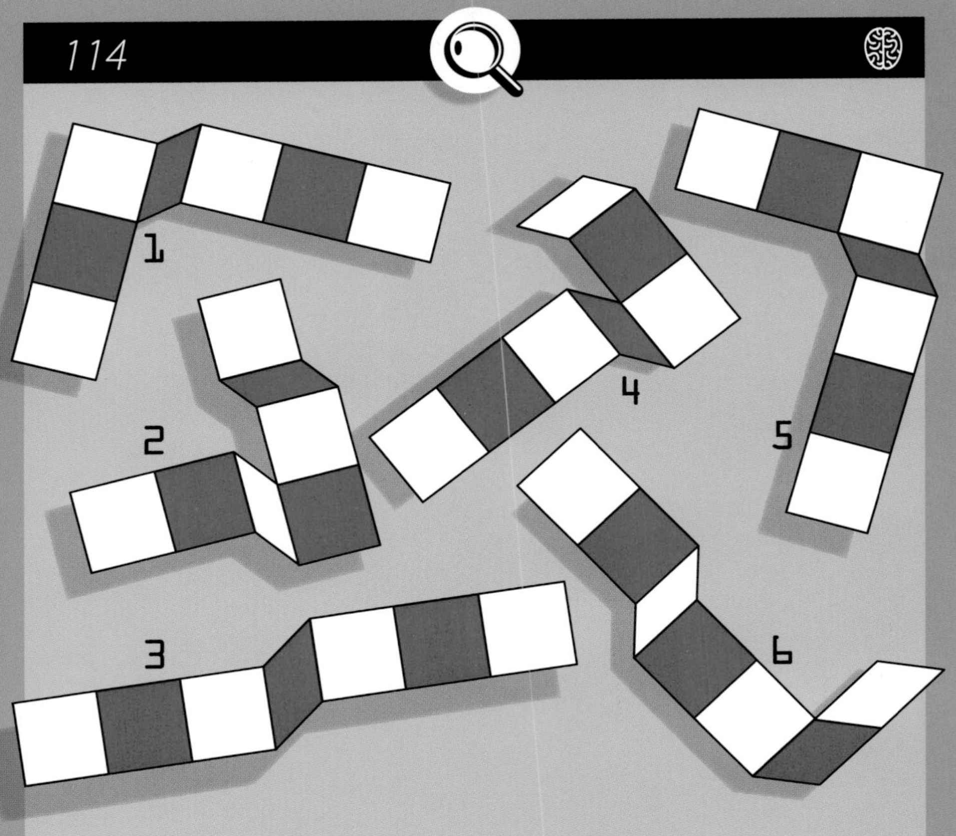

Which strip of folded paper (1–6) does not belong?

3-D Fun

115

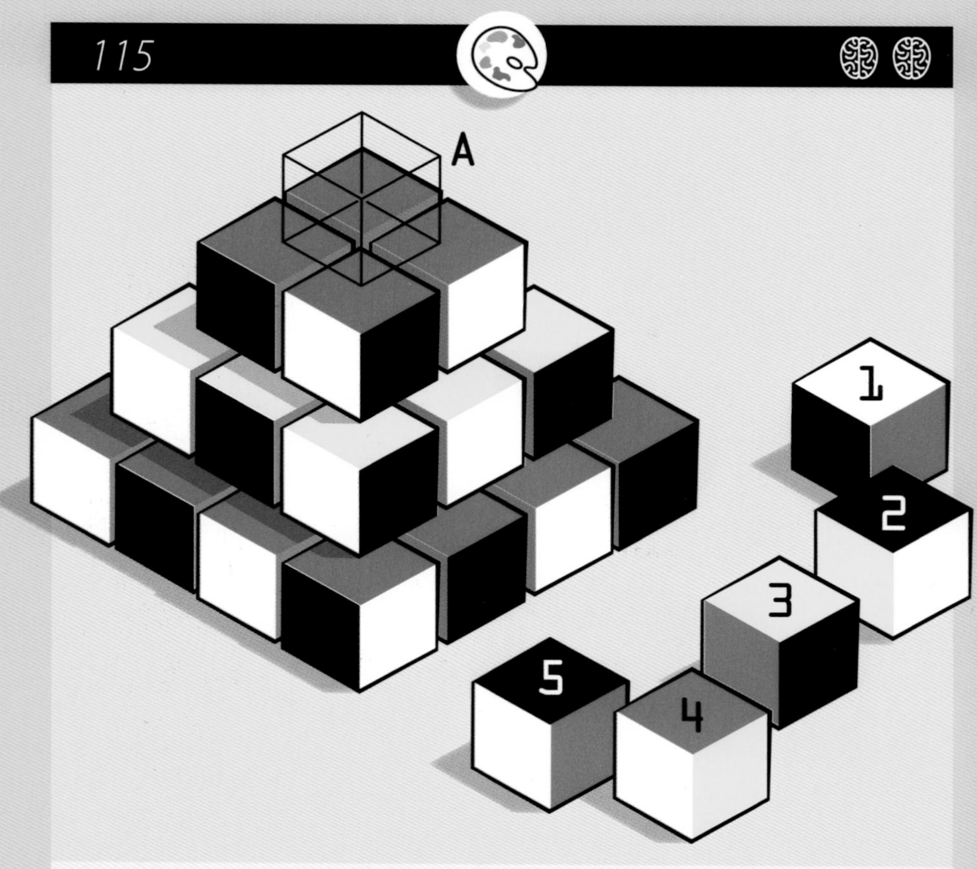

Which cube (1–5) fits on top of the pyramid?

3-D Fun

116

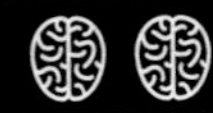

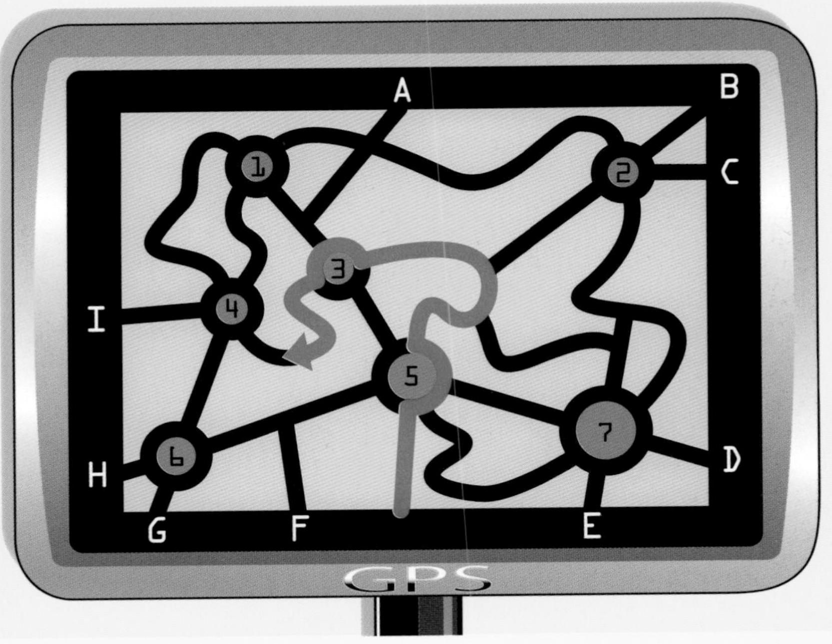

Where (A–I) will the GPS have the driver leave the map?

3-D Fun

117

Which number should replace the question mark?

3-D Fun

118

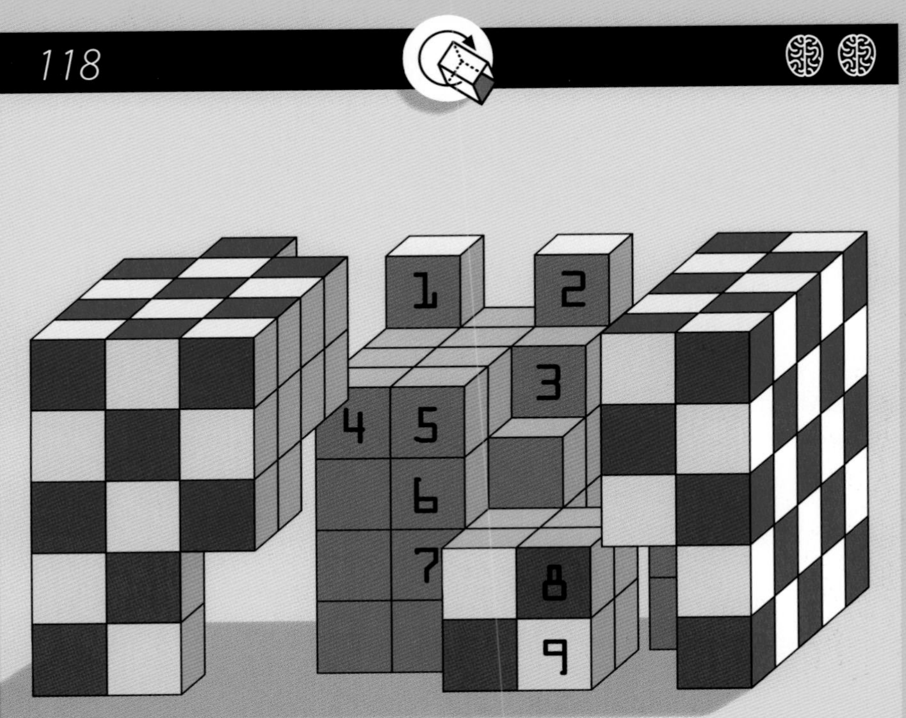

Which block (1–9) prevents one from being able to combine the three large parts into a completely filled cube?

3-D Fun

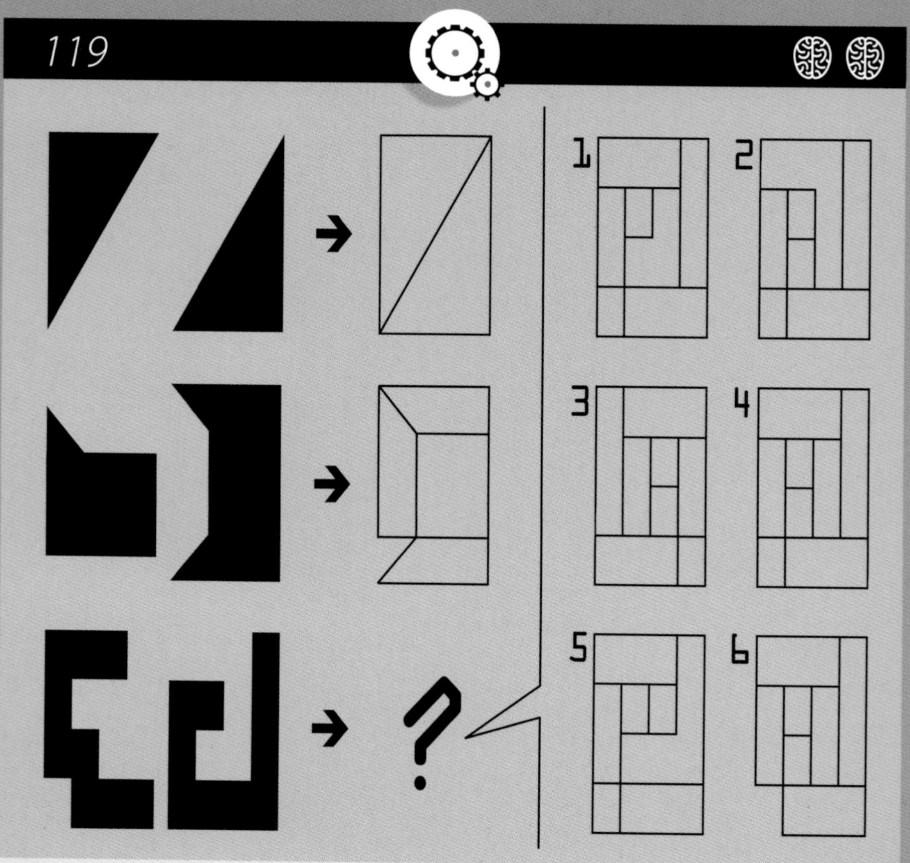

Which shape (1–6) should replace the question mark?

3-D Fun

120

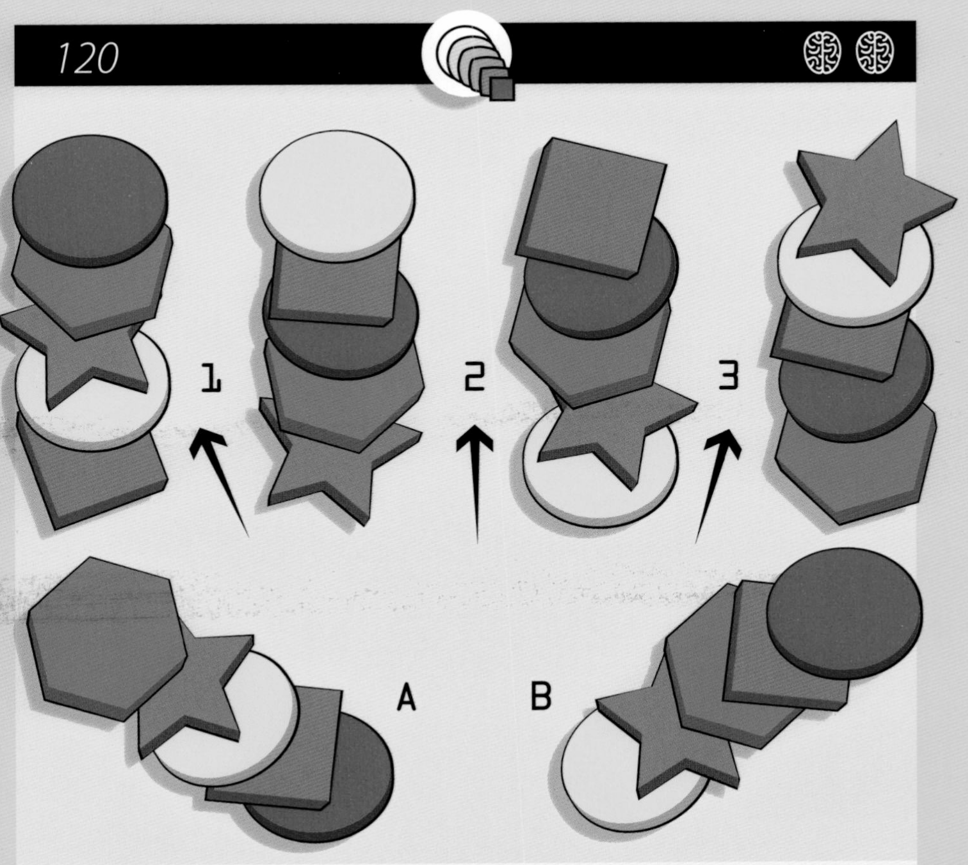

Only one stack (A or B) can be placed (1-3).
Which stack, and where?

Answer like this: B1.

3-D Fun

121

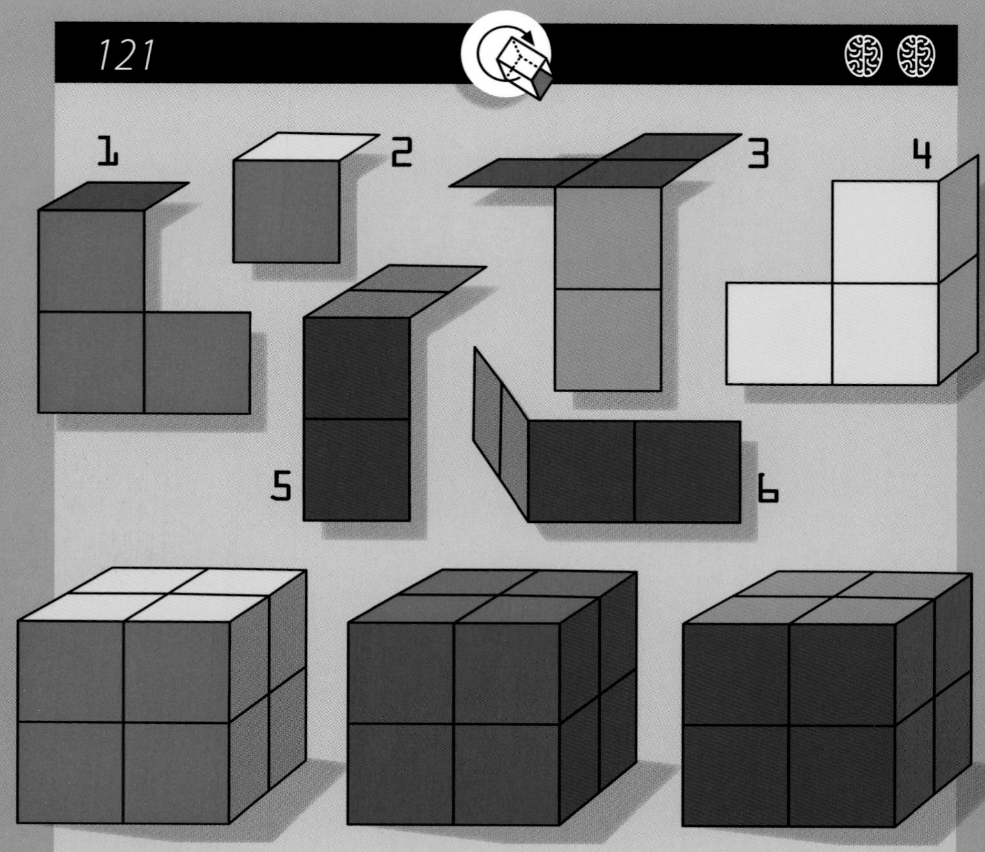

The same cube is shown from three different angles. You can complete the cube with the six pieces.

Which piece (1–6) is not useful?

3-D Fun

122

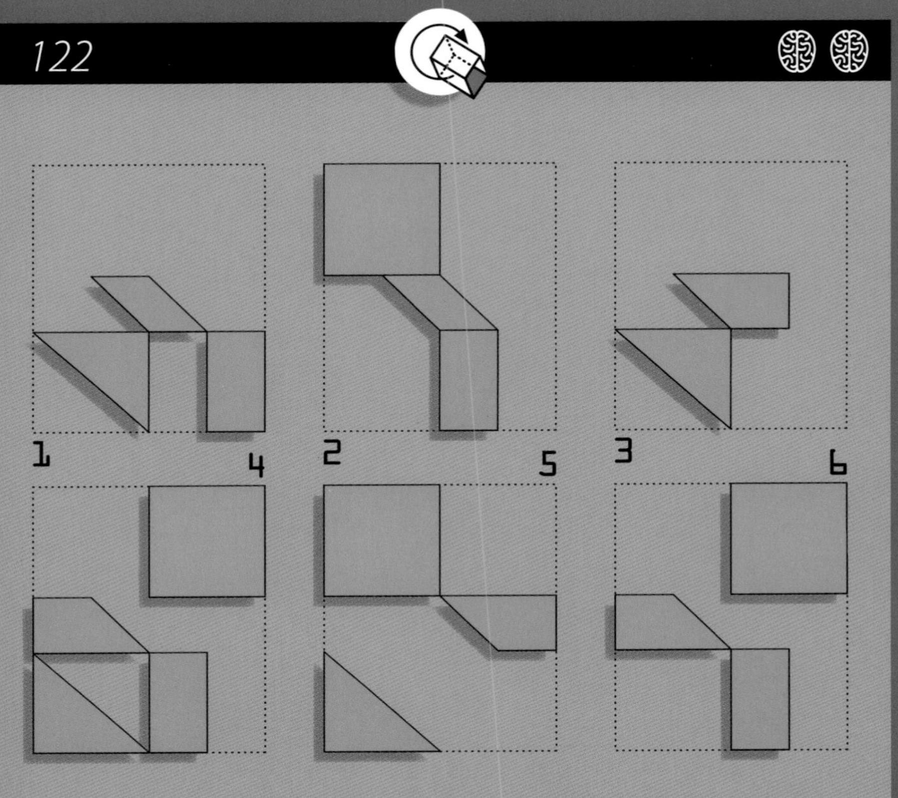

Which three puzzle pieces (1–6) fit together perfectly to create a rectangle?

Mind-Twisting Exercises

Divide a square into three identical triangles.

Solution on page 288

Eyes look, brains see

The eye works like a camera. It has a complete optical system, an aperture, and photosensitive film. The lens can zoom in and out and images are sharp on the retina at every distance. The retina consists of 125 million visual cells, cones, and rods that enable us to focus, distinguish colors, and detect details.

The physical image is sent through the optic nerve to more than thirty higher visual areas in the brain where the most important processing occurs. Scientists are still in the dark as to exactly what happens in those areas.

We do not only see with our eyes. Our brain participates actively; it interprets the information stored by our eyes. That does not always work well. Sometimes things go wrong in the nanosecond that the eye sends a physical image to the brain. Sometimes nothing goes wrong, but the brain has to make choices from tons of stimuli.

You do not always actually see what you see. 'Mind-Twisting Exercises' are living proof of this. Sometimes you see impossible configurations or things that are not there.

That is amazing, isn't it?

Mind-Twisting Exercises

123 COLOR ILLUSION

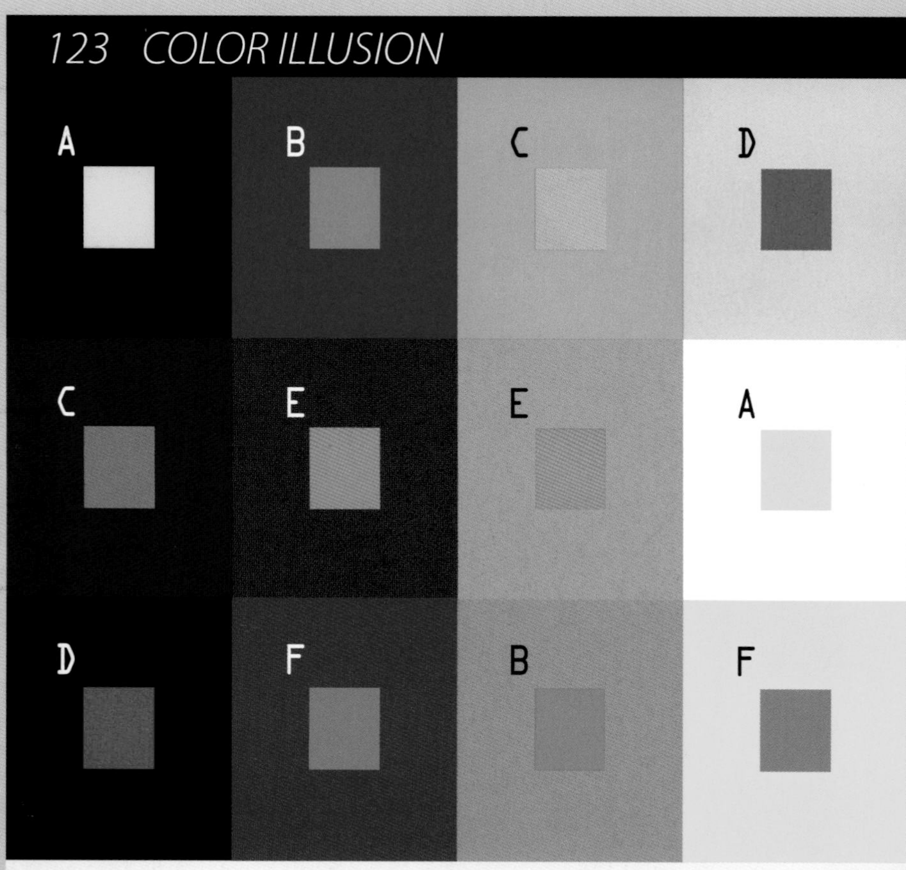

The two small squares with the same letter are the same color. For which letter (A–F) are the colors different?

Mind Twisting Exercises

124 WITH ONE LINE

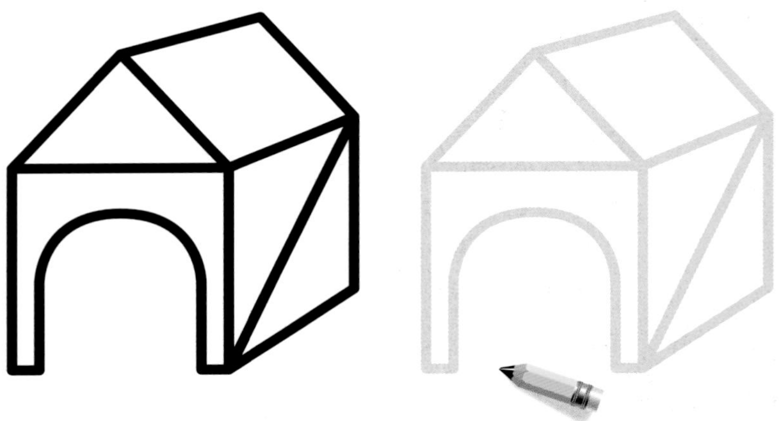

Try to draw this shape with one continuous line without lifting your pencil off the page and without any overlapping or crossing. There may be more than one solution.

Mind-Twisting Exercises

125 MIRROR IMAGE

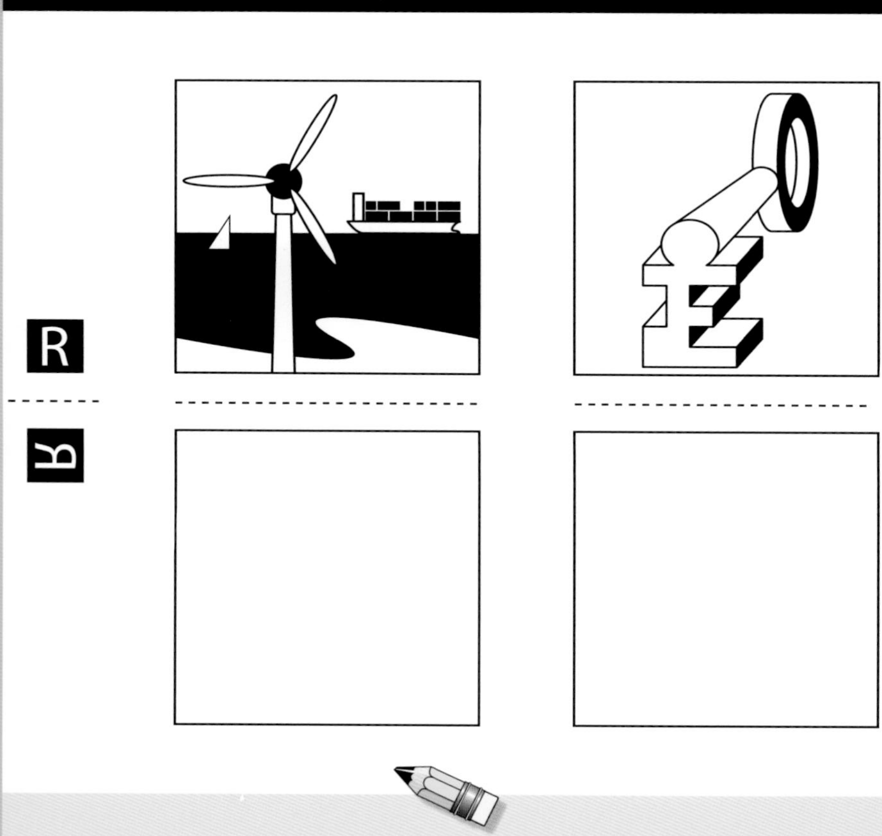

Draw the two pictures as if they were reflected and turned like the letter R.

Mind Twisting Exercises

126 CUT AND PASTE

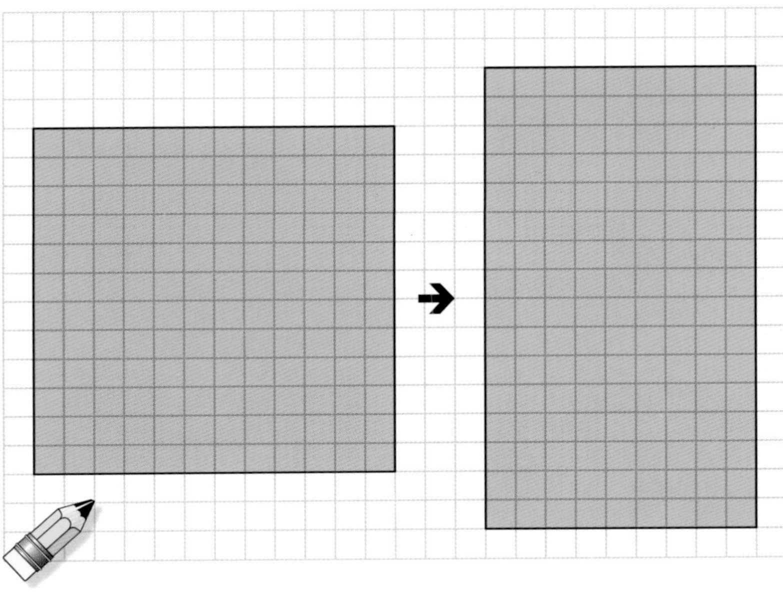

Cut the 12 by 12 figure along the grid lines into two pieces then paste them into a 9 by 16 rectangle.

Mind-Twisting Exercises

127 MAGIC SQUARE

On the engraving Melencolia I by Albrecht Dürer from 1514 there is a remarkable magic square because the rows, columns, and diagonals, as well as the four vertexes, the four middlemost numbers, the blocks of 2 by 2 numbers in the four corners, the two middlemost numbers in the first and last column, the two middlemost numbers in the upper and lowest row have the same sum, namely 34. Complete Dürer's magic square with the numbers 1 to 16 on a separate piece of paper.

128 MATCHSTICK PUZZLE

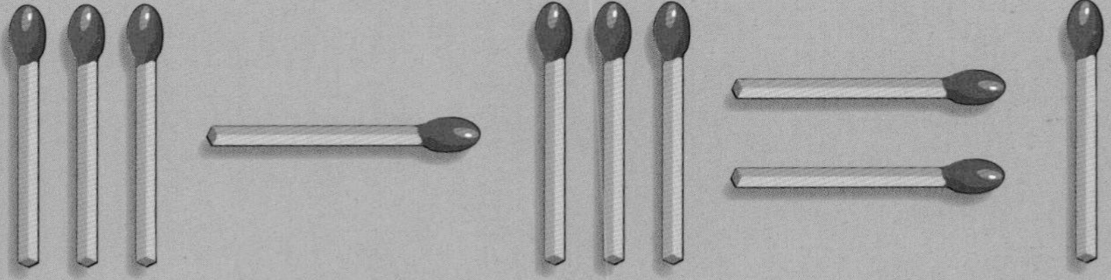

Move two matches to complete the formula.

Mind-Twisting Exercises

129 STEREOGRAM

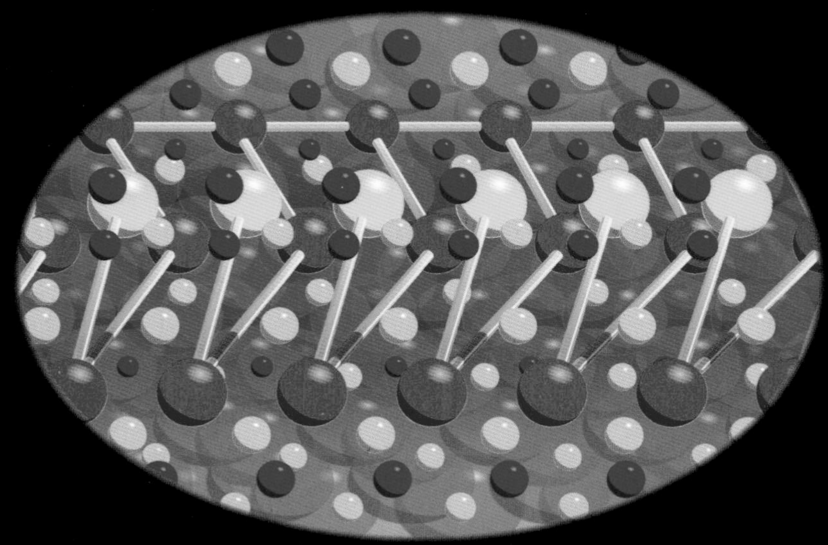

How to look at this floating atoms stereogram?

Make sure the stereogram is well-lit. To allow the eyes to focus behind the image (diverge), hold the stereogram in front of your face until you touch your nose. At that distance the brain cannot see a sharp image.
Next step: increase—very slowly—the distance between the stereogram and your eyes while suppressing the urge to focus. At a certain moment the brain will fixate on a pattern in which the distance between the elements coincides with the convergence angle of your two eyes.
Another way is to look behind the stereogram until the correct convergence angle is found.

Good luck!

Mind Twisting Exercises

130 ILLUSION

5ƐƐIN9

I5

I3Ɛ∠IƐ?IN9

What letter should replace the question mark?

Mind-Twisting Exercises

131 ILLUSION

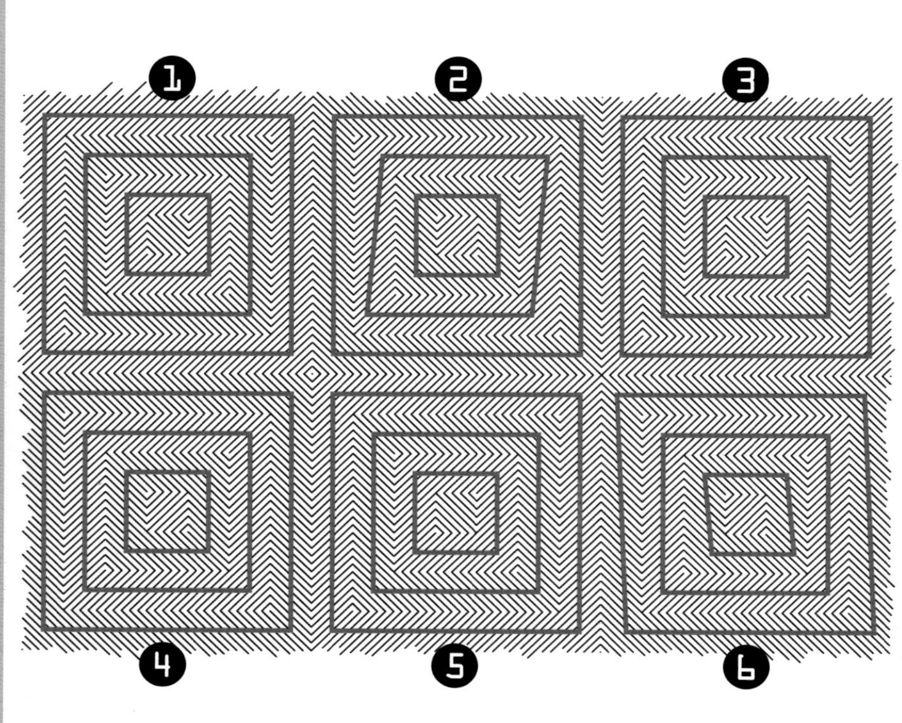

Which groups (1–6) with three squares also have parallelograms?

Mind Twisting Exercises

132 TRIPLETS

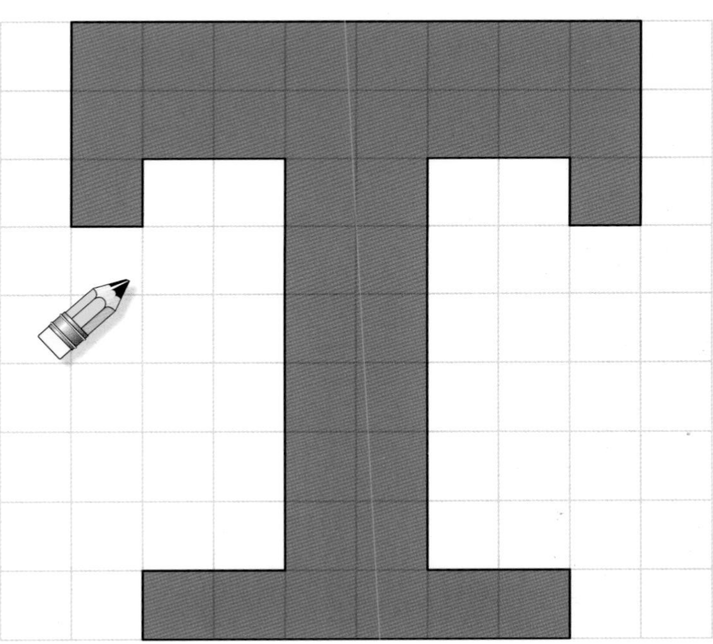

Cut the T-figure along the grid lines into three identical pieces.

Mind-Twisting Exercises

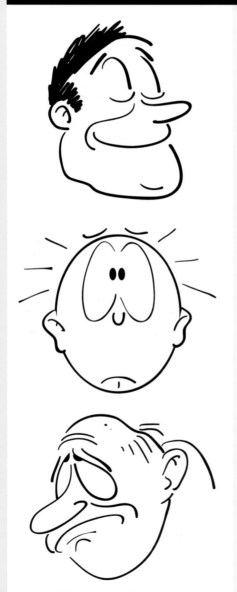

133 EMOTIONS

1

....
....
....
....
....

2

....
....
....
....
....

3

....
....
....
....
....

Mind Twisting Exercises

134 EMOTIONS

4

....
....
....
....
....

5

....
....
....
....
....

Look for at least five emotions for every character.

Every emotion must be different.

Mind-Twisting Exercises

135 WITH ONE LINE

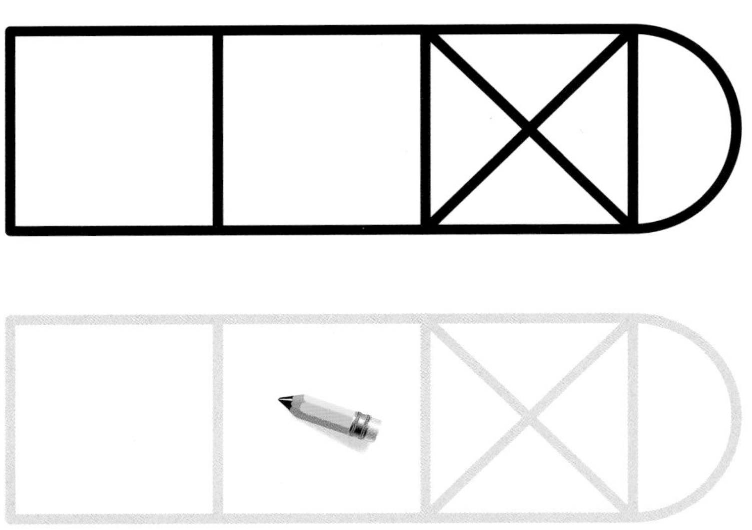

Try to draw this shape with one continuous line without lifting your pencil off the page and without any overlapping or crossing. There may be more than one solution.

Mind Twisting Exercises

136 ILLUSION

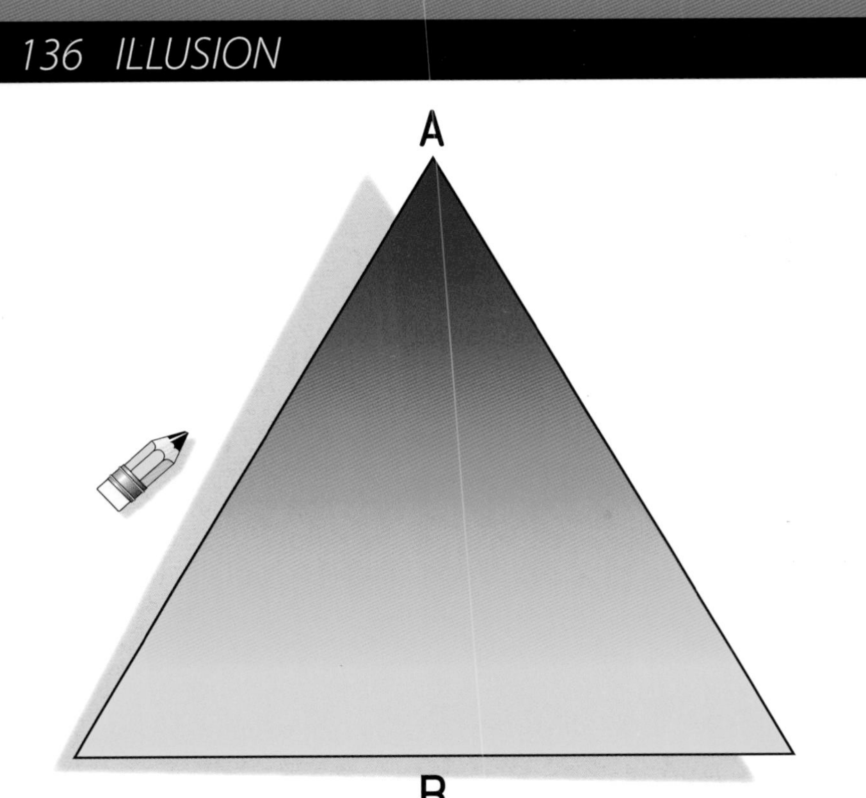

Try to indicate on this triangle where the middle is between top A and base B.

Do this as quickly as possible, then measure your results.

Mind-Twisting Exercises

137 Prime numbers

2

3

5

7

11

13

17

19

?

A prime number is a natural number greater than one that is only divisible by one and itself.

Prime numbers are an important subject in mathematics because they are used when securing digital information.

What is the next prime number in the series?

Mind Twisting Exercises

138 MATCHSTICK PUZZLE

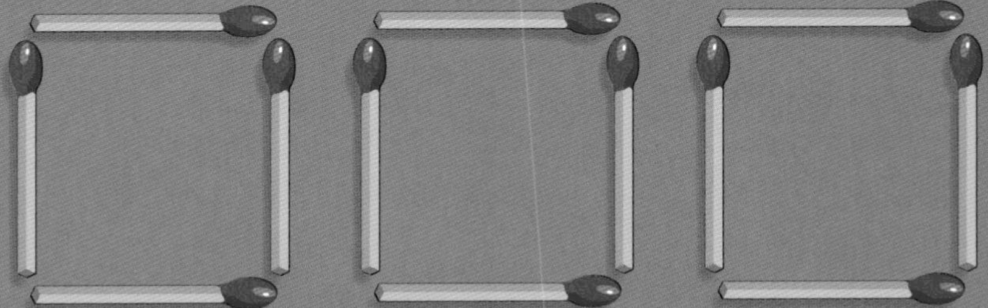

Move two matches to create 4:30 a.m.

Mind-Twisting Exercises

139 WHAT A DIFFERENCE A DAY MAKES

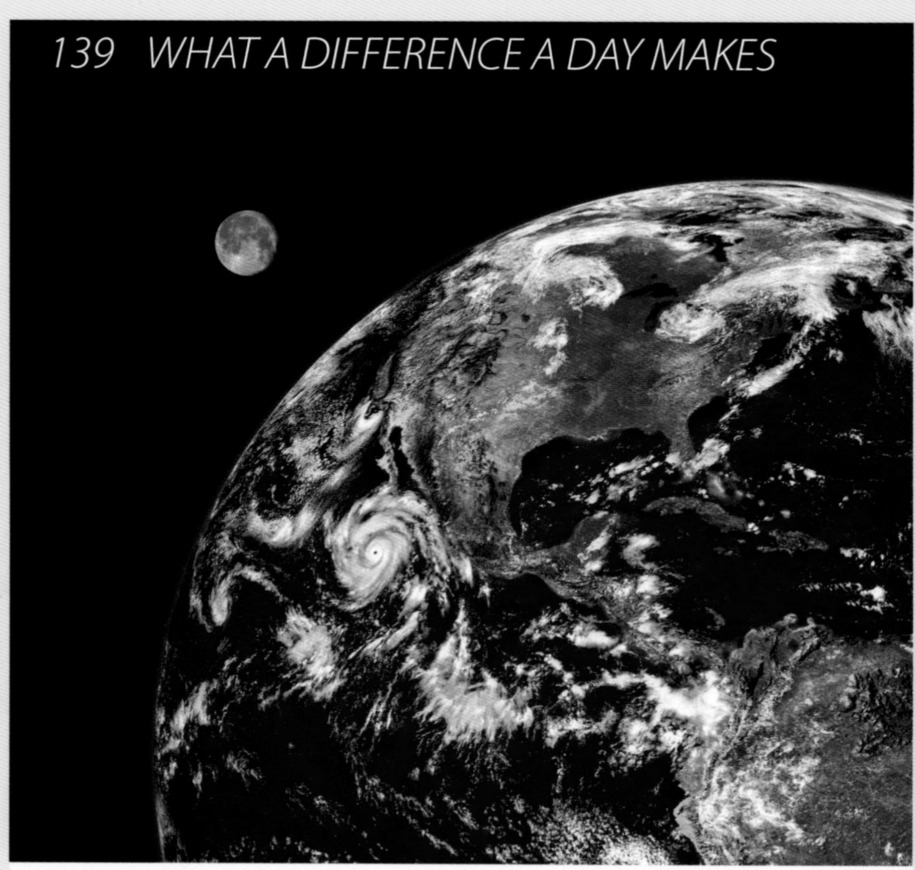

What is one day after the day after the day before yesterday?

Mind Twisting Exercises

140 THE FRANKFURTER ILLUSION

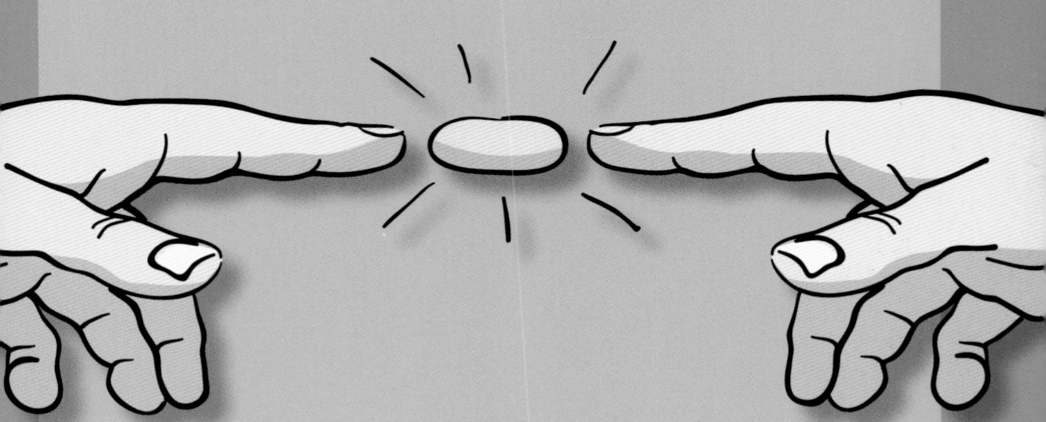

Place your index fingers one inch from each other, about ten inches in front of your eyes, and look at an object on the horizon between the opening.

Mind-Twisting Exercises

141 NUMBER PYRAMID

Each stone lists the sum of the two stones under it. The stones on the bottom row contain all the numbers from 0 to 7 one time.

Complete the pyramid based on the six numbers that are already filled in.

Mind Twisting Exercises

142 CUT AND PASTE

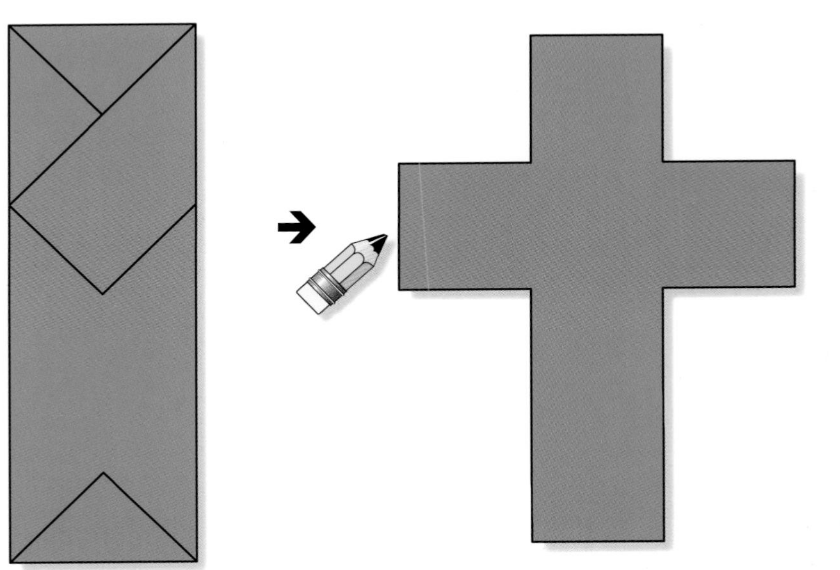

Use the five puzzle pieces in the rectangle to build the cross next to it. The puzzle pieces must not overlap.

Mind-Twisting Exercises

143 FIBONACCI SERIES

0

1

1

2

3

5

8

13

?

The further this series develops, the more the relationship between two consecutive numbers approaches the golden section.

At the golden section the largest of the two parts (numbers) is to the smallest (number) as the whole line (sum of the numbers) is to the largest (number).

What is the next number in the Fibonacci series?

Mind Twisting Exercises

144 WITH ONE LINE

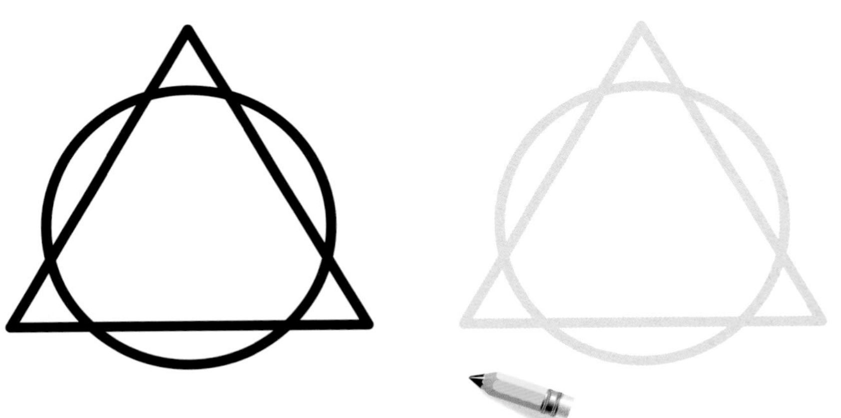

Try to draw this shape with one continuous line without lifting your pencil off the page and without any overlapping or crossing. There may be more than one solution.

Mind-Twisting Exercises

145 DISCOVER YOUR DOMINANT EYE

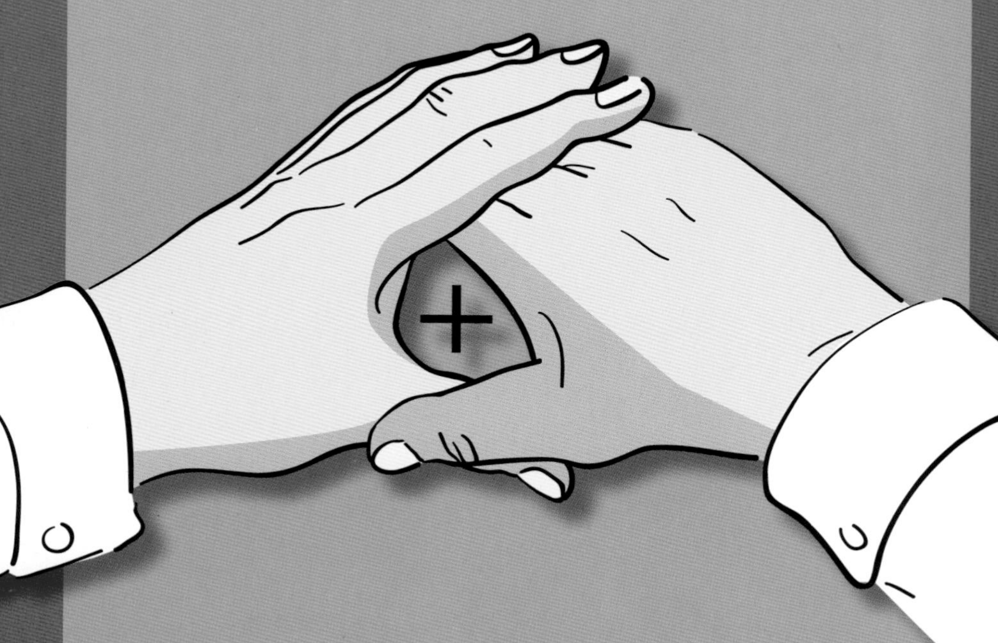

1 *At a distance of twenty inches, place your hands on top of each other, creating a peephole.*
2 *Look through the hole with both eyes at an object in the background.*
3 *Close your left eye and then your right without moving your head and hands. The eye that continues to see the object is the dominant eye. It's the eye that you automatically use to look through the viewfinder of a camera.*

Mind Twisting Exercises

146 MATCHSTICK PUZZLE

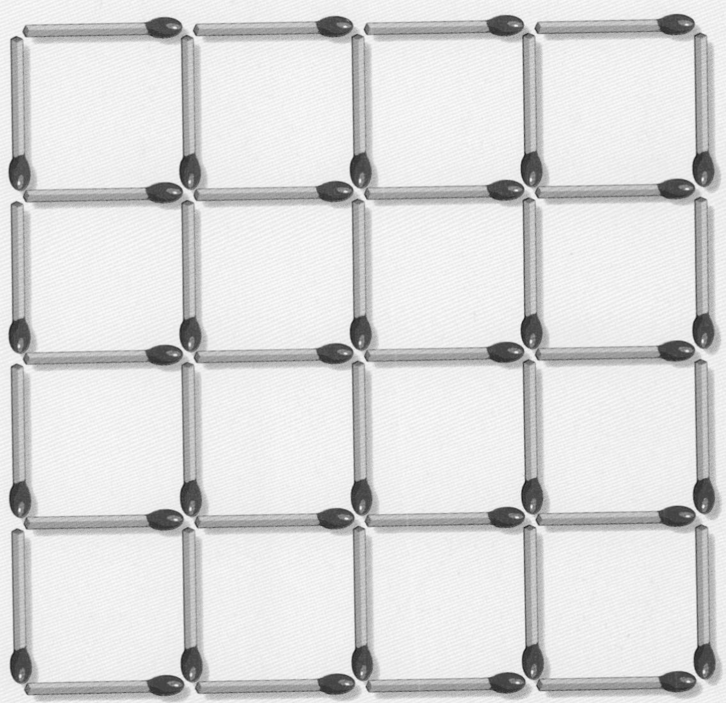

Remove nine matches so that there are no more squares.

GRID PUZZLES

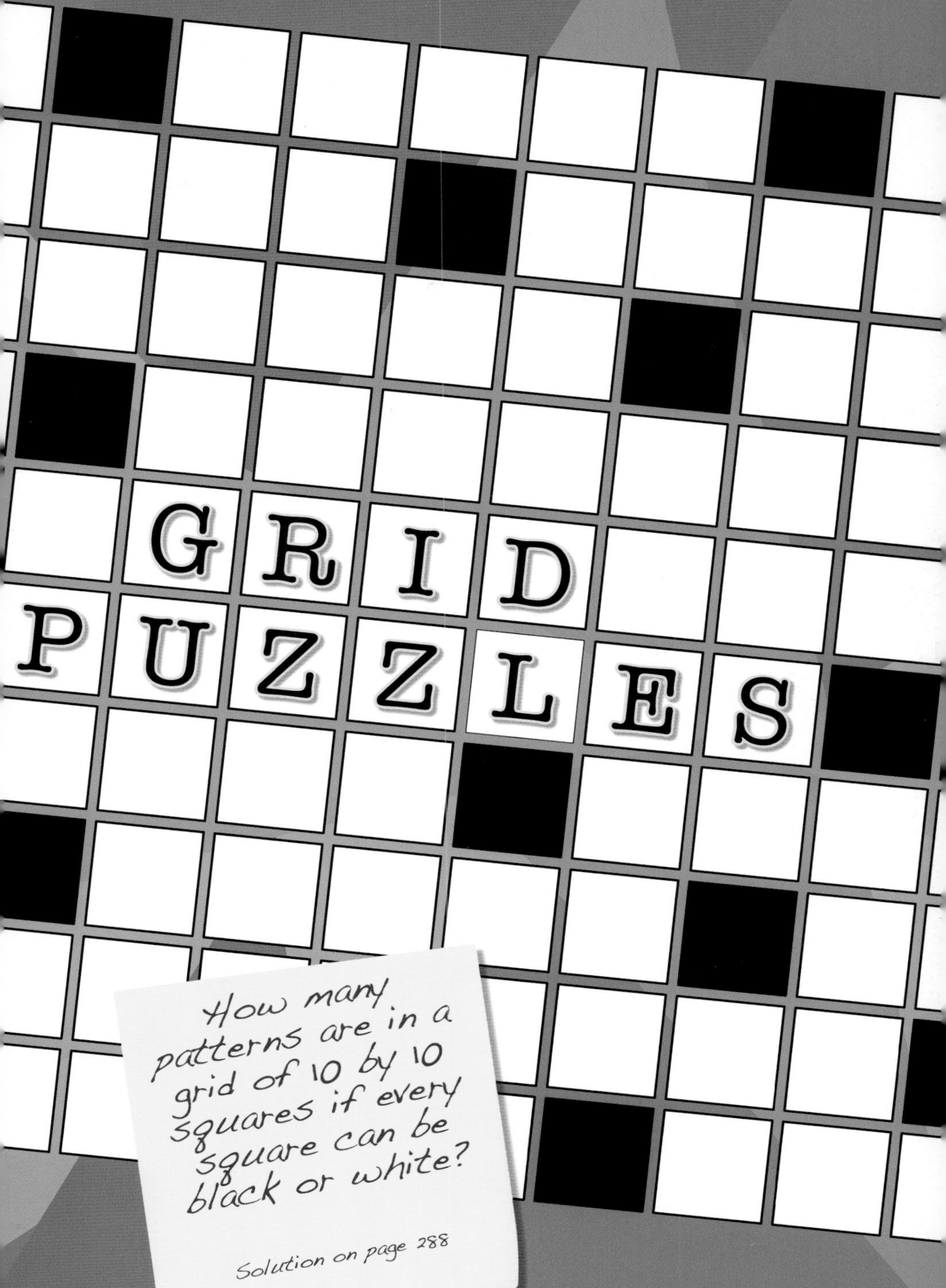

How many patterns are in a grid of 10 by 10 squares if every square can be black or white?
Solution on page 288

Quick Four-Letter Word Paper Game

1. All players draw the same grid of **4** by **4** squares.
2. Toss a coin to see who starts calling out a letter in turn.
3. Letters can only be used **4** times.
4. As soon as a letter is called out, all players must use the letter in their grids; they can choose which square to write the letter in.
5. When the grid is full, players count how many **4**-letter words they can make. The words can be horizontal, vertical, or diagonal, from bottom to top, from left to right, and vice-versa.
6. The player with the most **4**-letter words is the winner.
7. You should decide in advance which words do not count.

P	R	Y	S
A	A	C	X
N	O	R	I
T	D	A	K

Words in grid

PANT

PARK

IRON

SCOT

KRAP (percussion instrument)

Grid Puzzles

DIRECTIONS NUMBER CLUSTERS

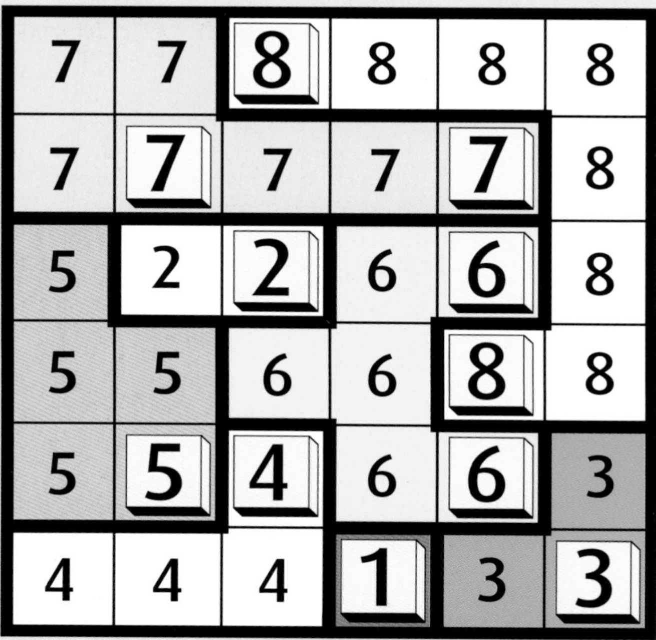

Complete the grid by constituting adjoining clusters that consist of as many cubes as the number on the cubes. At cube 5, for instance, you will have to make a five cube cluster. Two or more figure cubes of the same value belong to the same cluster. You can only place your cubes along horizontal and/or vertical lines.

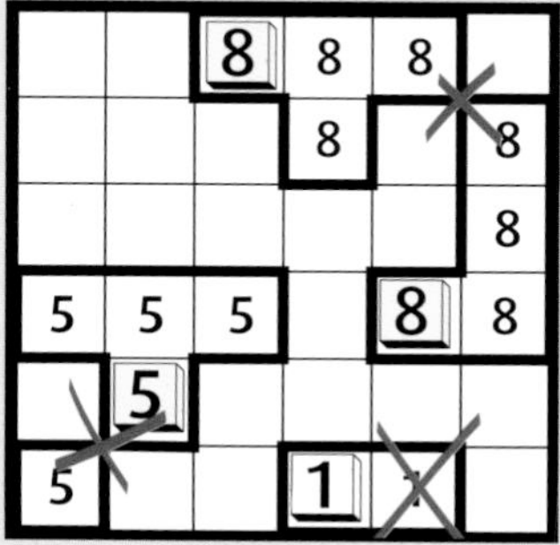

Grid Puzzles

147 NUMBER CLUSTERS

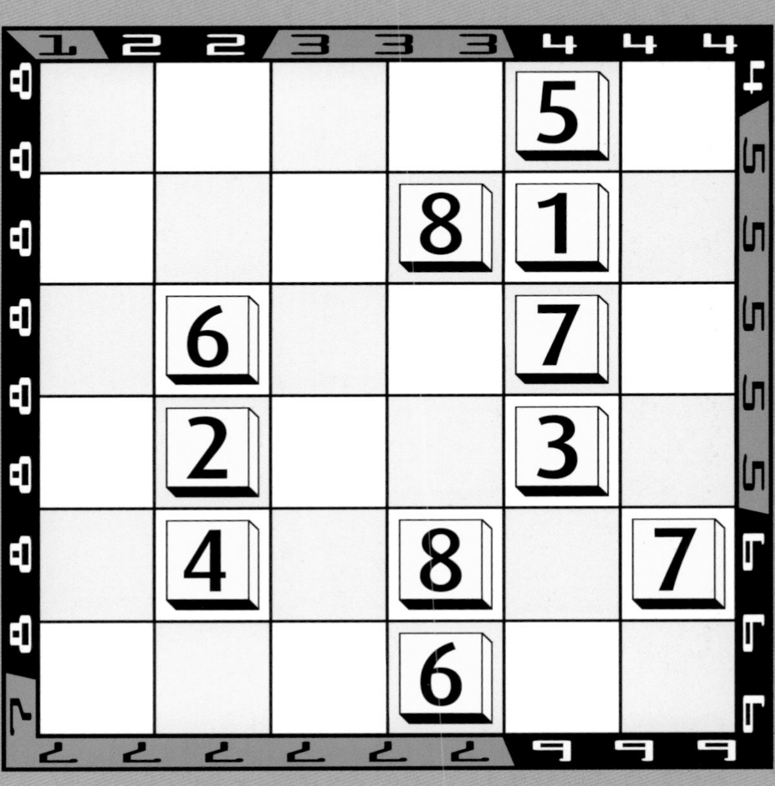

TEST ZONE

Grid Puzzles

148 NUMBER CLUSTERS

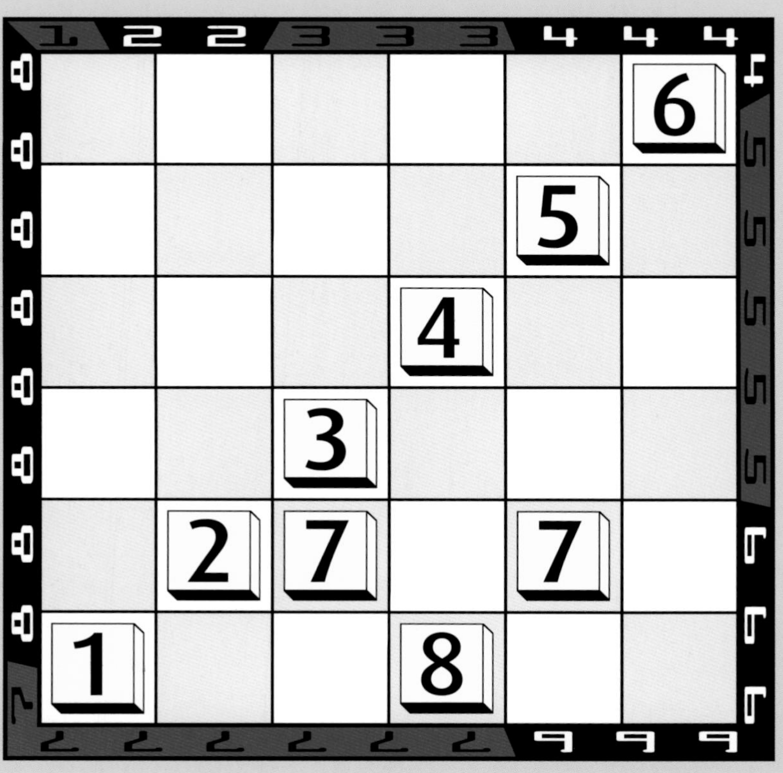

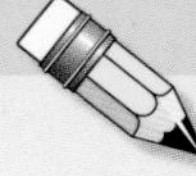

TEST ZONE

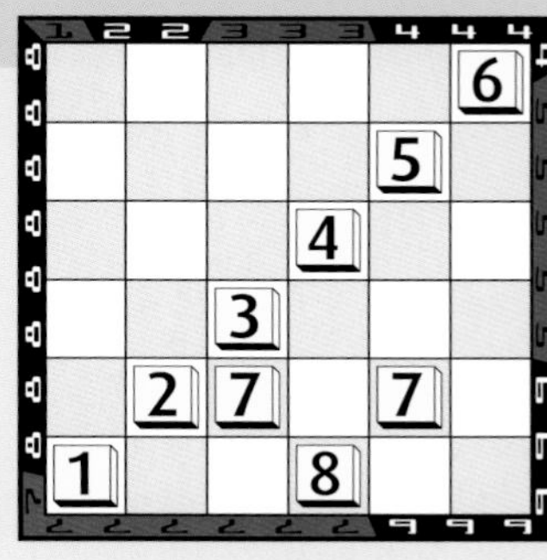

Grid Puzzles

149 NUMBER CLUSTERS

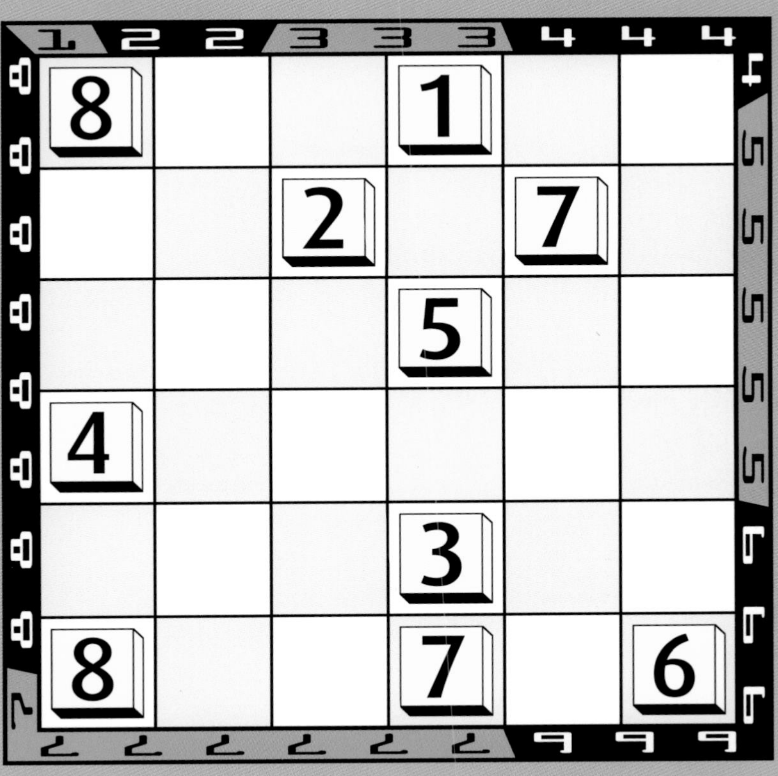

TEST ZONE

8			1		
		2		7	
			5		
4					
			3		
8			7		6

Grid Puzzles

150 NUMBER CLUSTERS

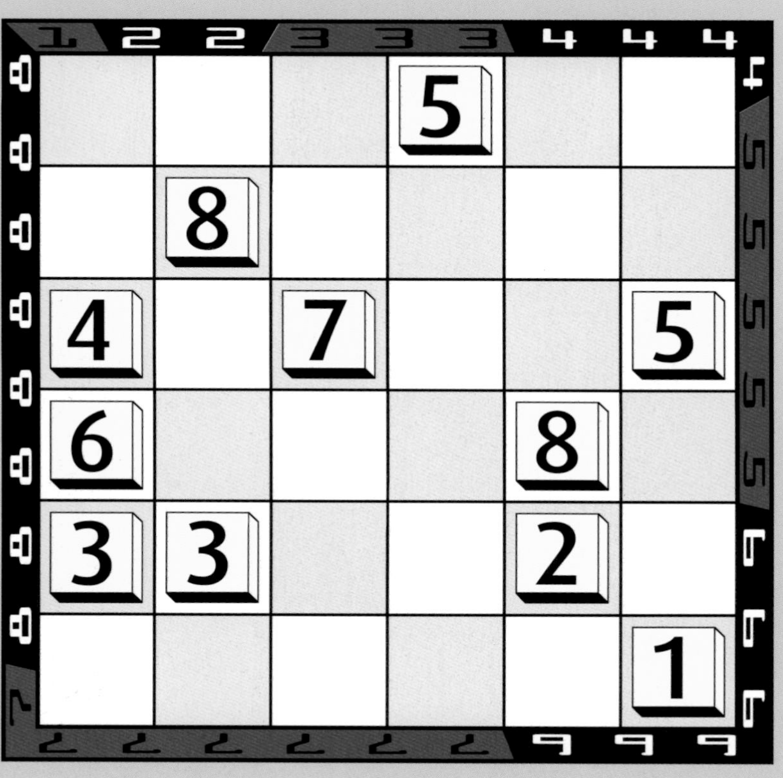

TEST ZONE

Grid Puzzles

151 NUMBER CLUSTERS

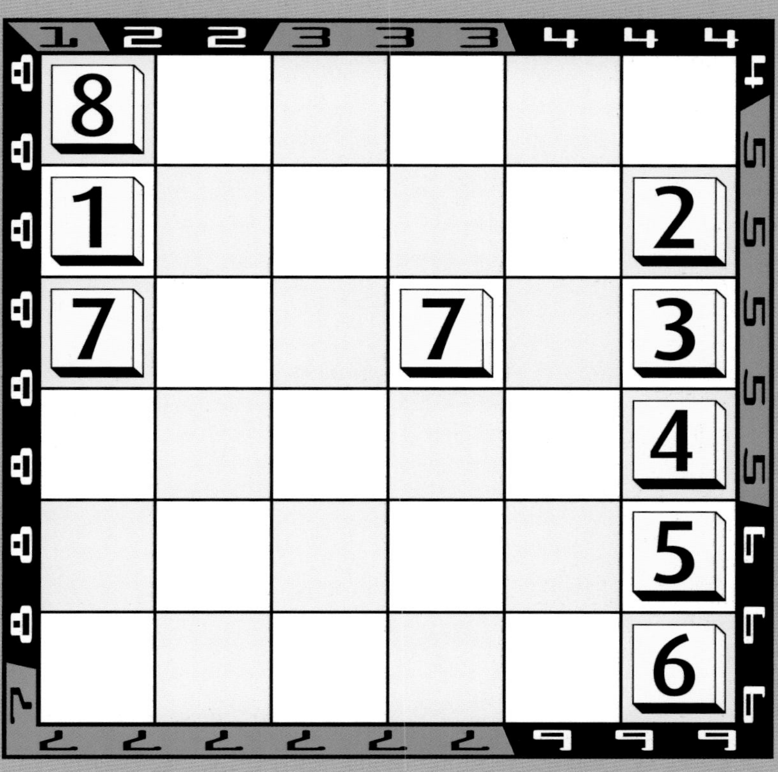

TEST ZONE

Grid Puzzles

152 NUMBER CLUSTERS

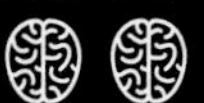

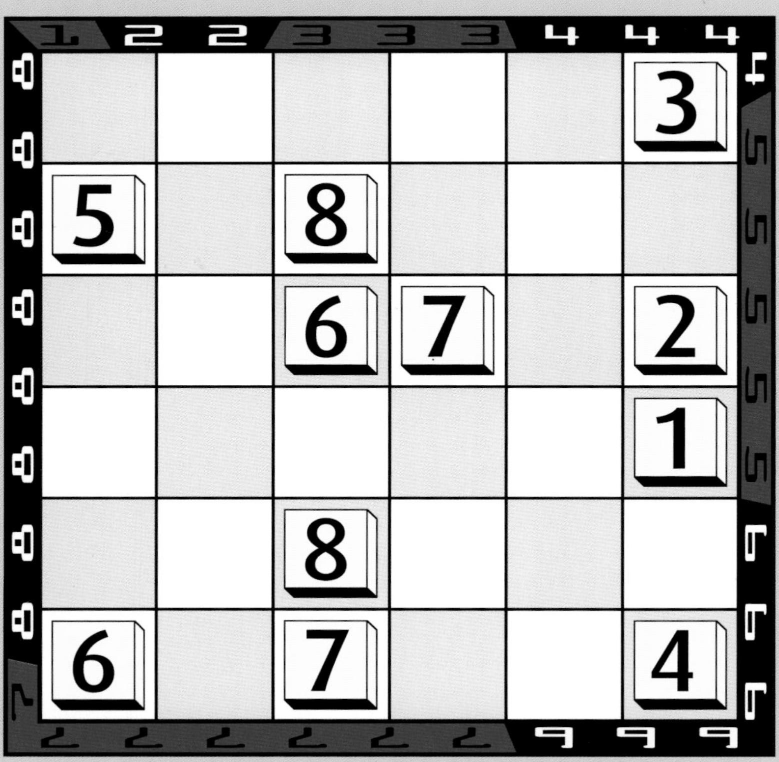

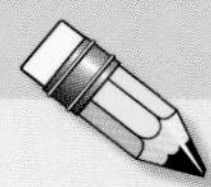

TEST ZONE

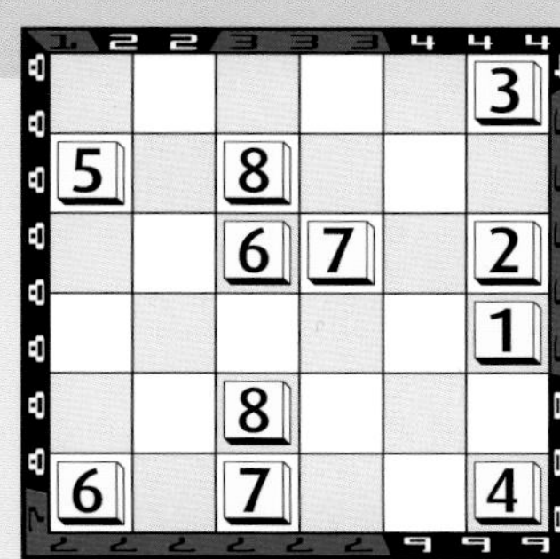

Grid Puzzles

153 NUMBER CLUSTERS

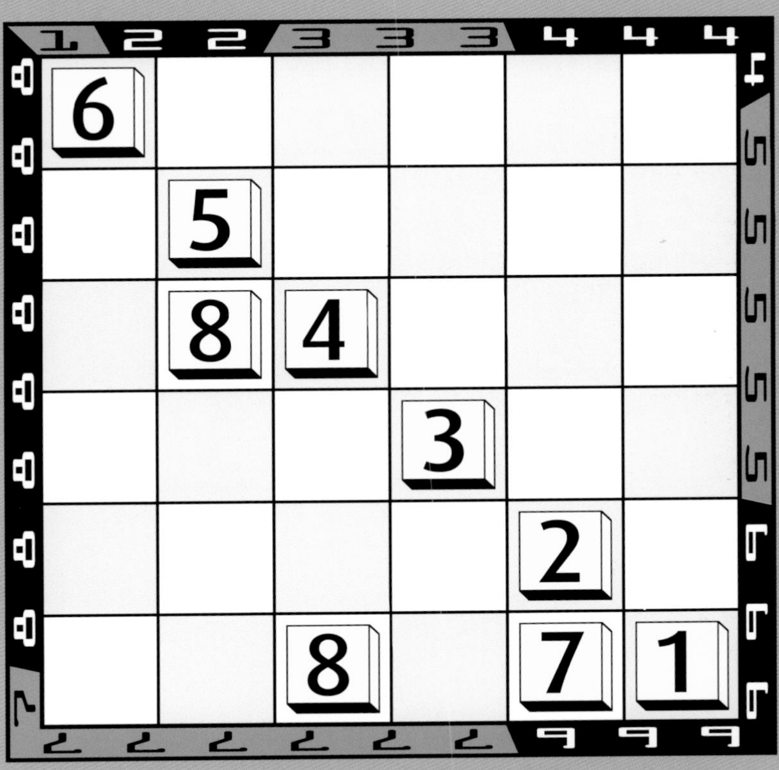

TEST ZONE

Grid Puzzles

154 NUMBER CLUSTERS

		7	6		5
	8				
	2		1	3	
	4				

TEST ZONE

		7	6		5
	8				
	2		1	3	
	4				

Grid Puzzles

155 NUMBER CLUSTERS

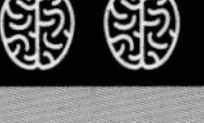

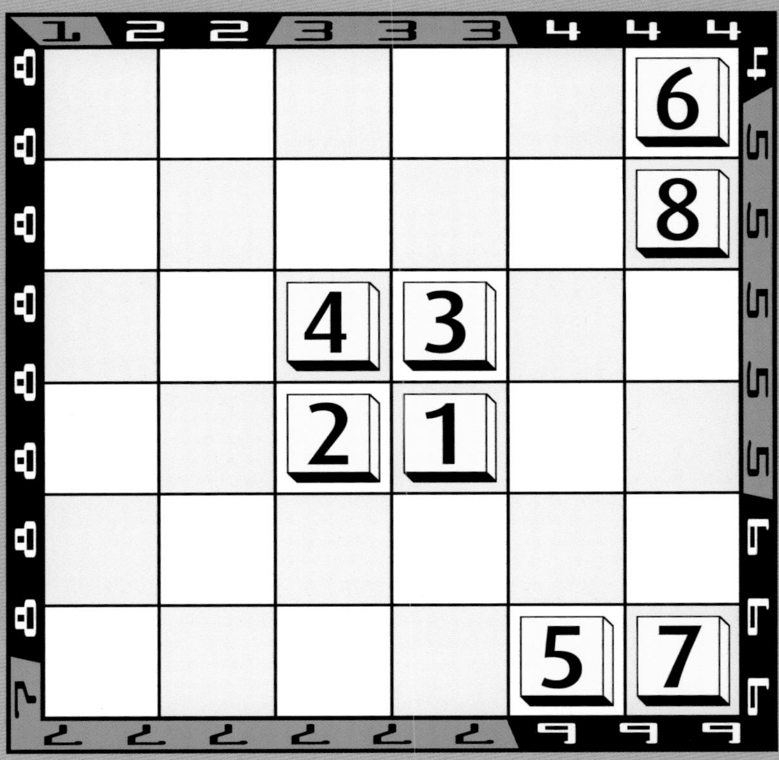

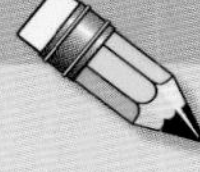

TEST ZONE

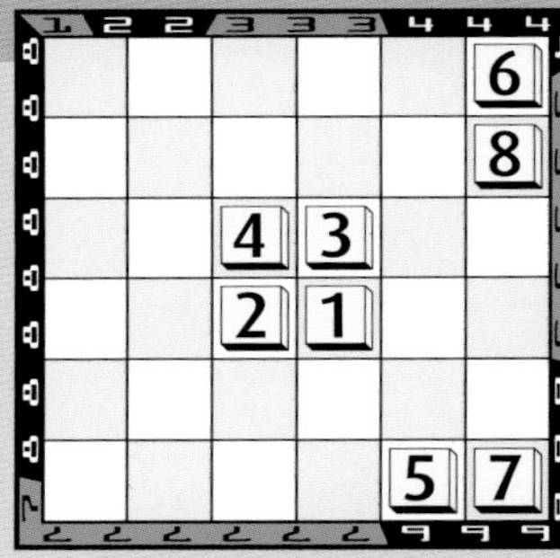

Grid Puzzles

156 NUMBER CLUSTERS

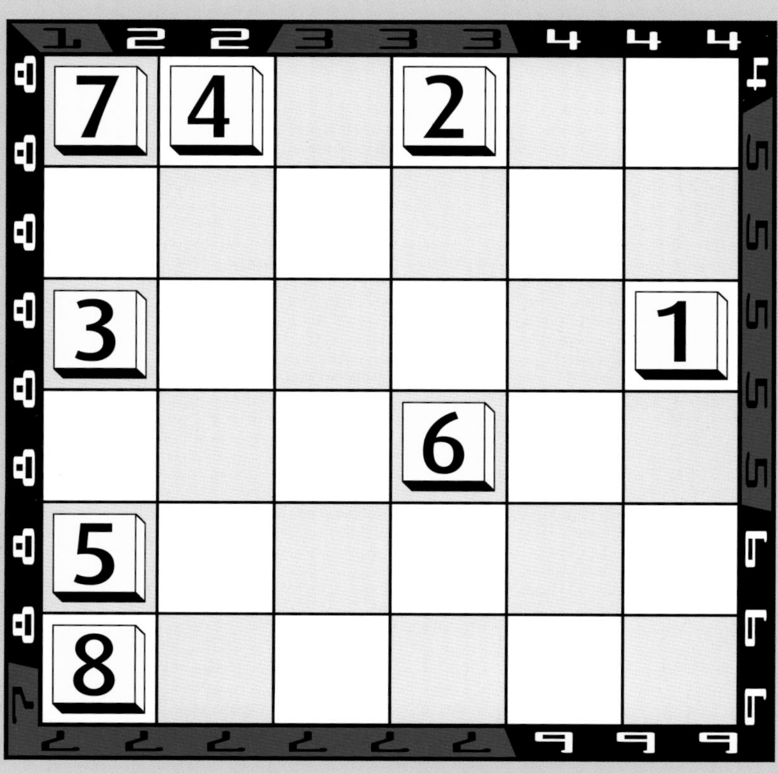

TEST ZONE

Grid Puzzles

157 NUMBER CLUSTERS

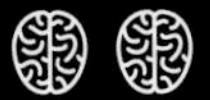

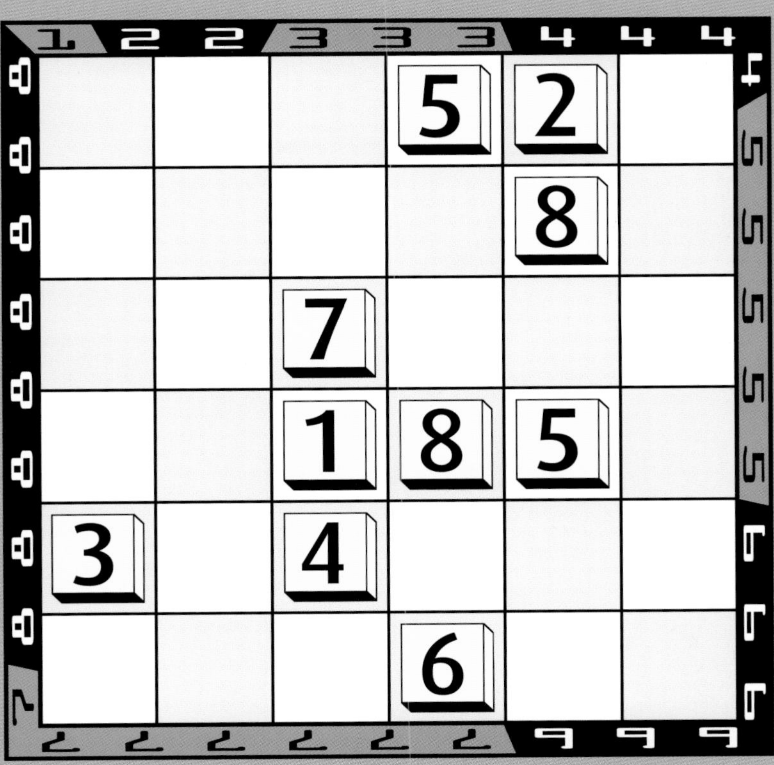

TEST ZONE

Grid Puzzles

DIRECTIONS SUNNY WEATHER FORECAST

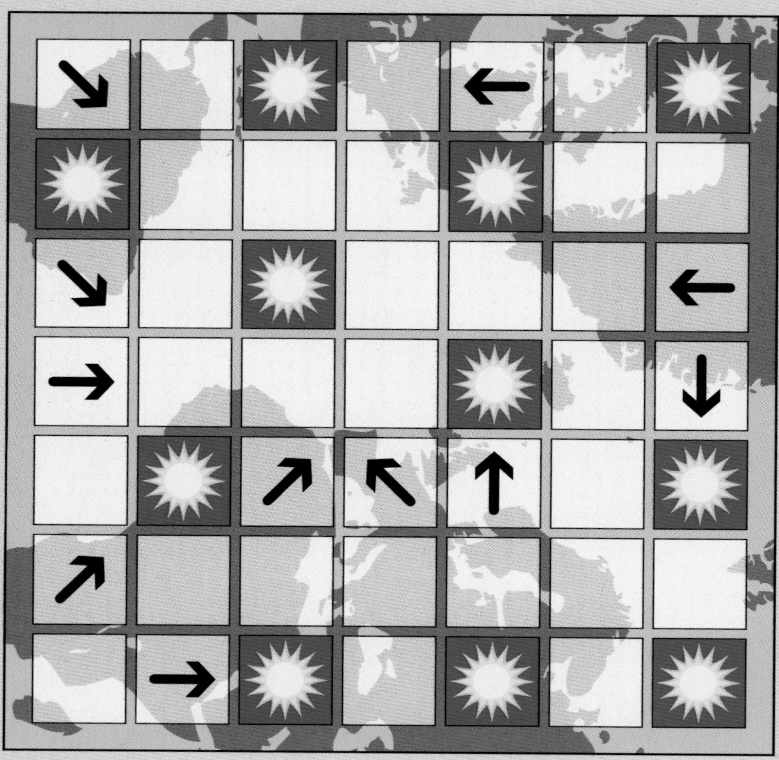

1 *Knowing that every arrow points to a sun and that no sun can touch another vertically, horizontally, or diagonally, find the missing suns.*

2 *A symbol cannot be located on an arrow.*

3 *We show one symbol.*

Grid Puzzles

158 SUNNY WEATHER FORECAST

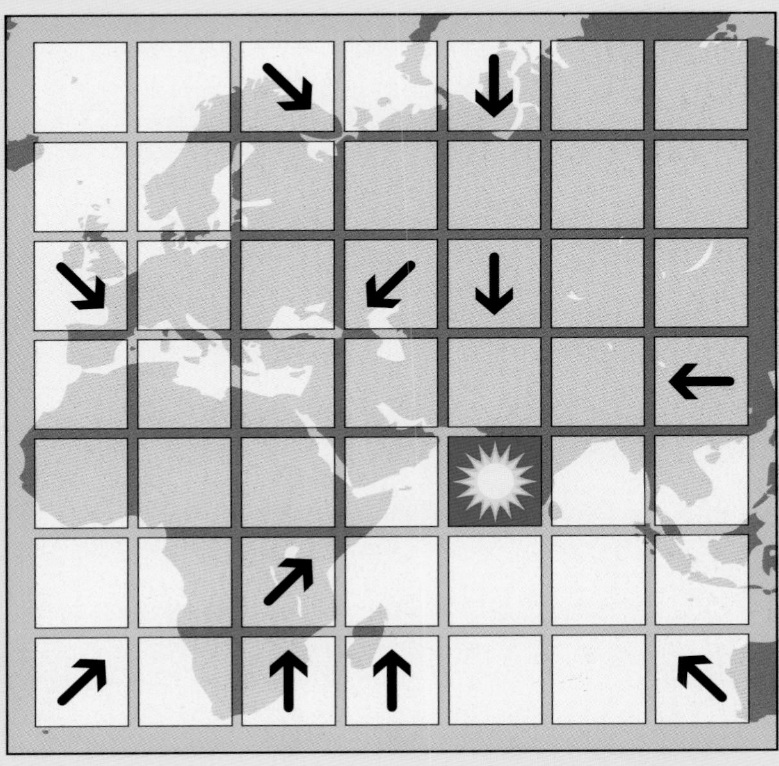

TEST ZONE

Find eleven sunny areas.

Grid Puzzles

159 SUNNY WEATHER FORECAST

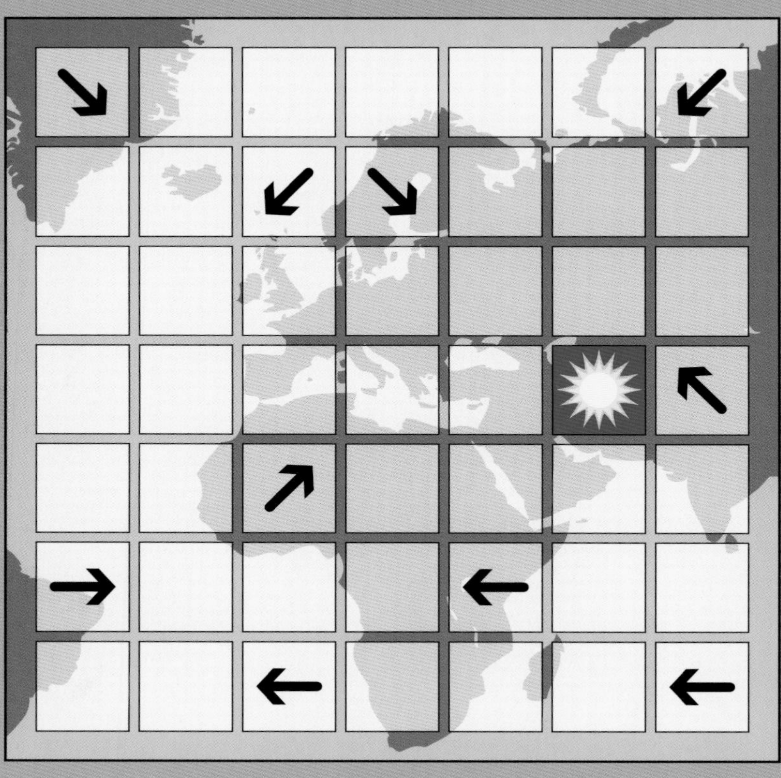

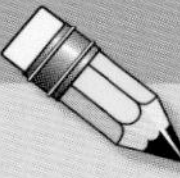

TEST ZONE

Find ten sunny areas.

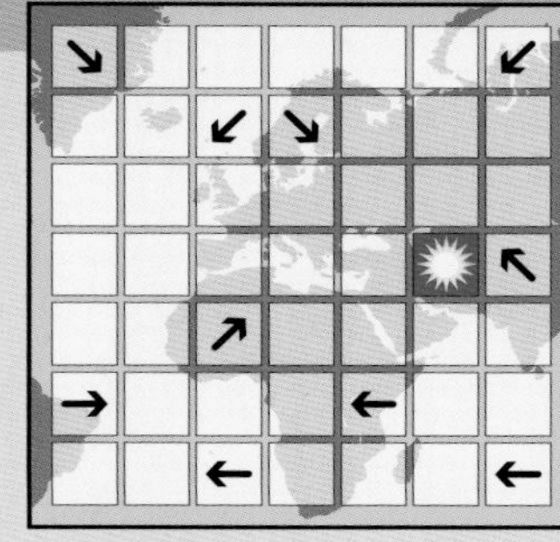

Grid Puzzles

160 SUNNY WEATHER FORECAST

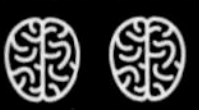

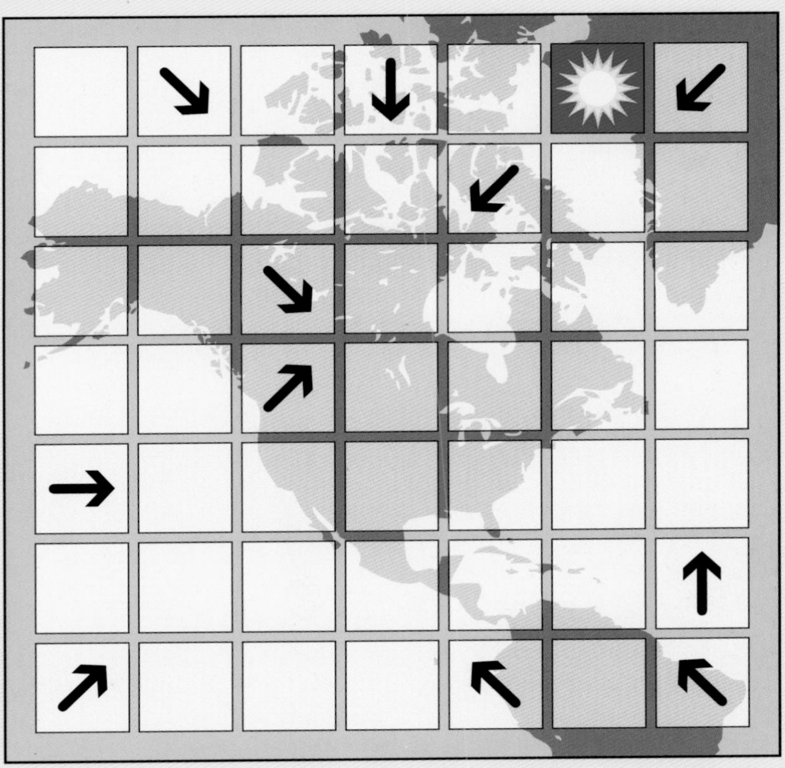

TEST ZONE

Find eleven sunny areas.

Grid Puzzles

161 SUNNY WEATHER FORECAST

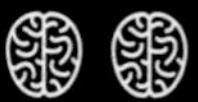

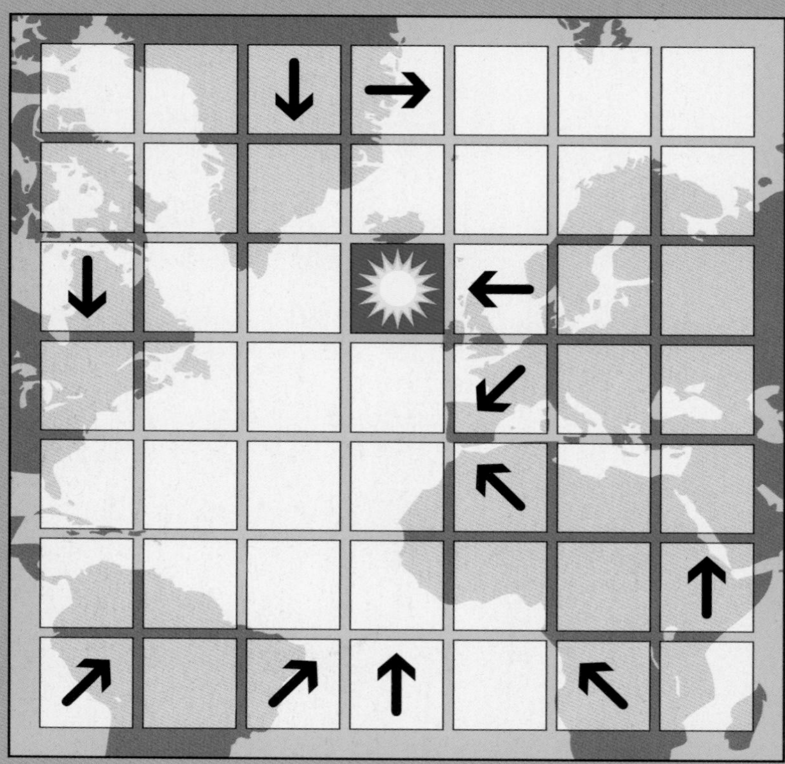

TEST ZONE

Find eleven sunny areas.

Grid Puzzles

162 SUNNY WEATHER FORECAST

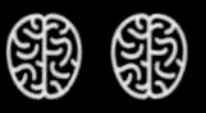

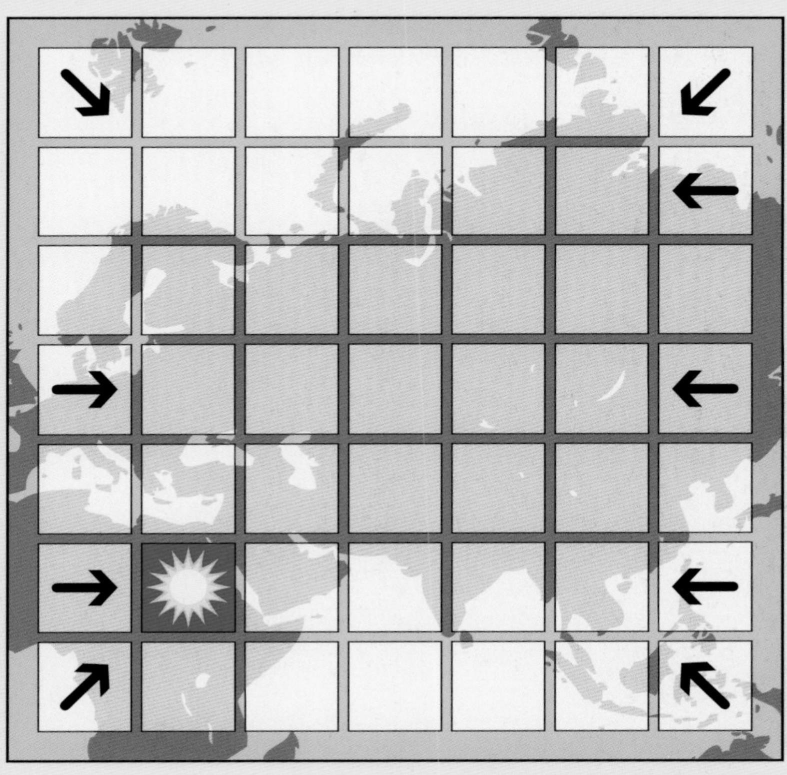

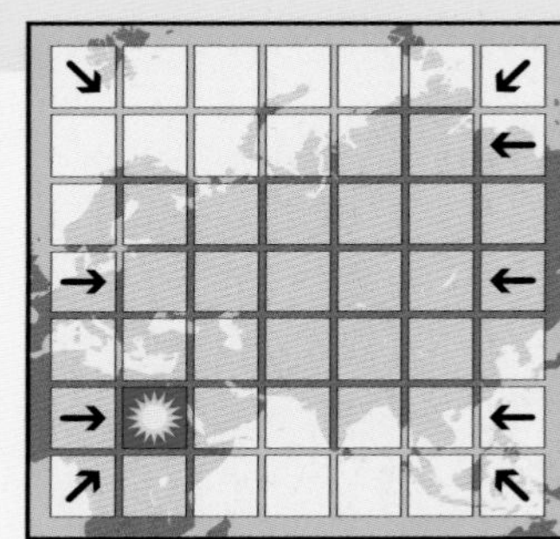

TEST ZONE

Find nine sunny areas.

Grid Puzzles

163 SUNNY WEATHER FORECAST

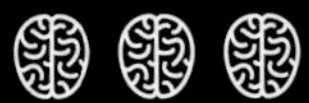

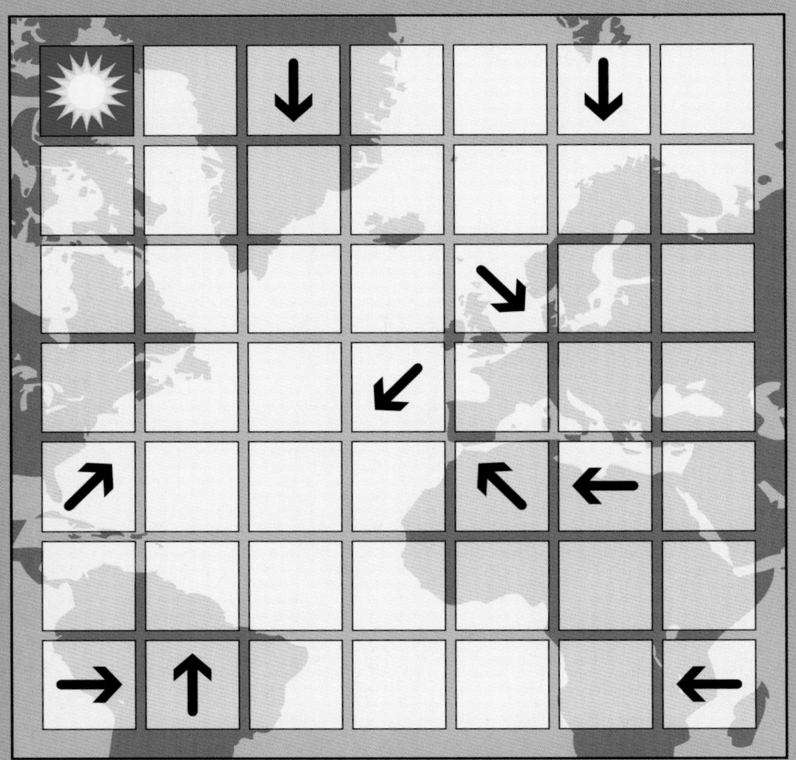

TEST ZONE

Find ten sunny areas.

Grid Puzzles

164 SUNNY WEATHER FORECAST

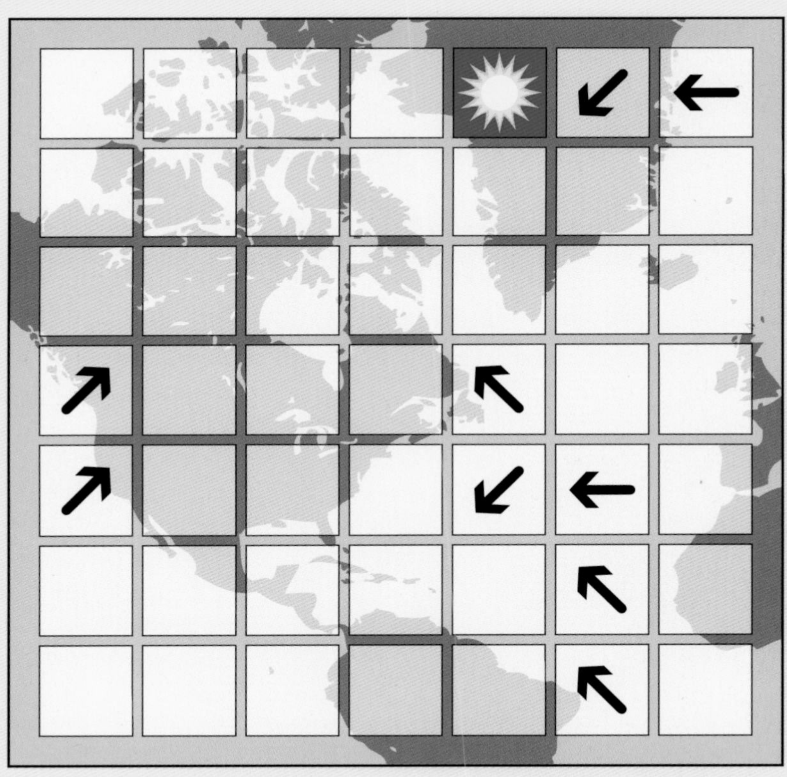

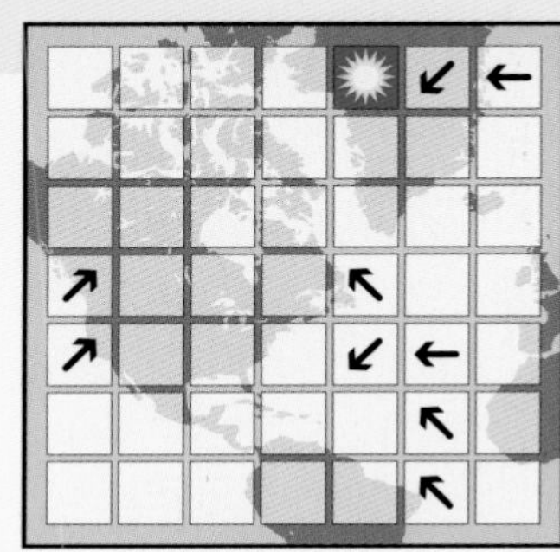

TEST ZONE

Find nine sunny areas.

Grid Puzzles

165 SUNNY WEATHER FORECAST

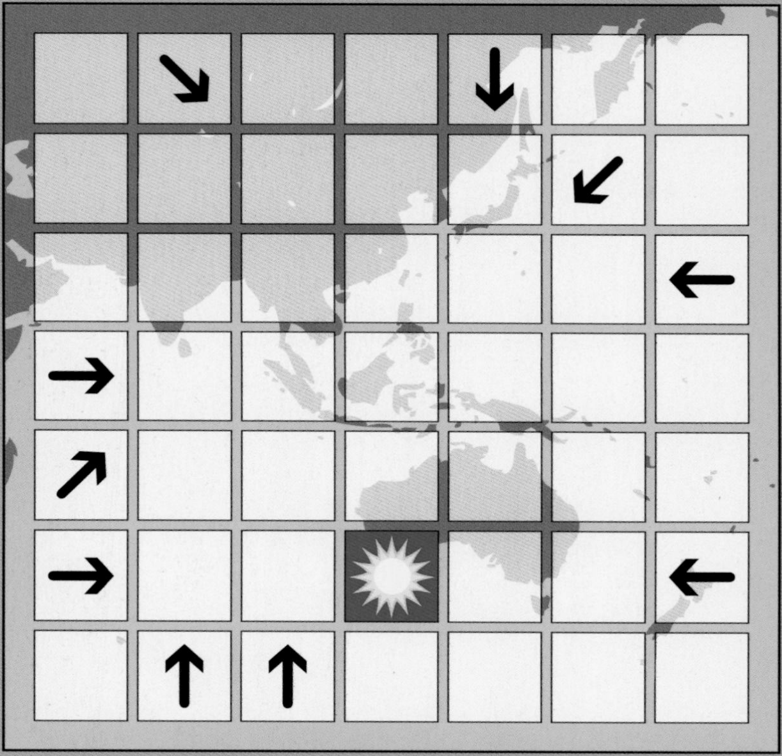

	↘			↓		
					↙	
						←
→						
↗						
→			☀			←
	↑	↑				

TEST ZONE

Find ten sunny areas.

Grid Puzzles

166 SUNNY WEATHER FORECAST

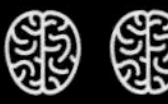

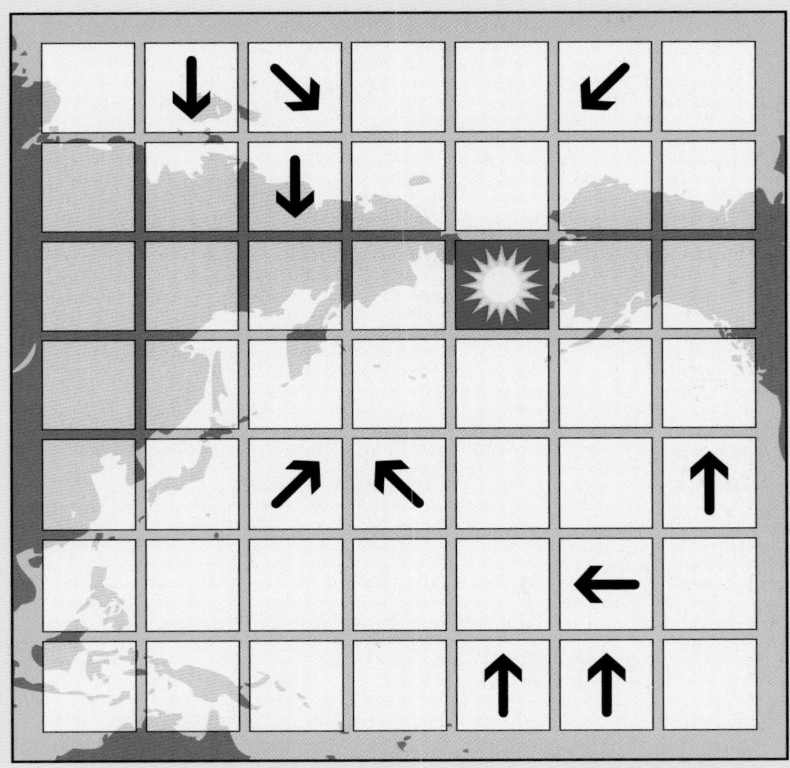

TEST ZONE

Find ten sunny areas.

Grid Puzzles

167 SUNNY WEATHER FORECAST

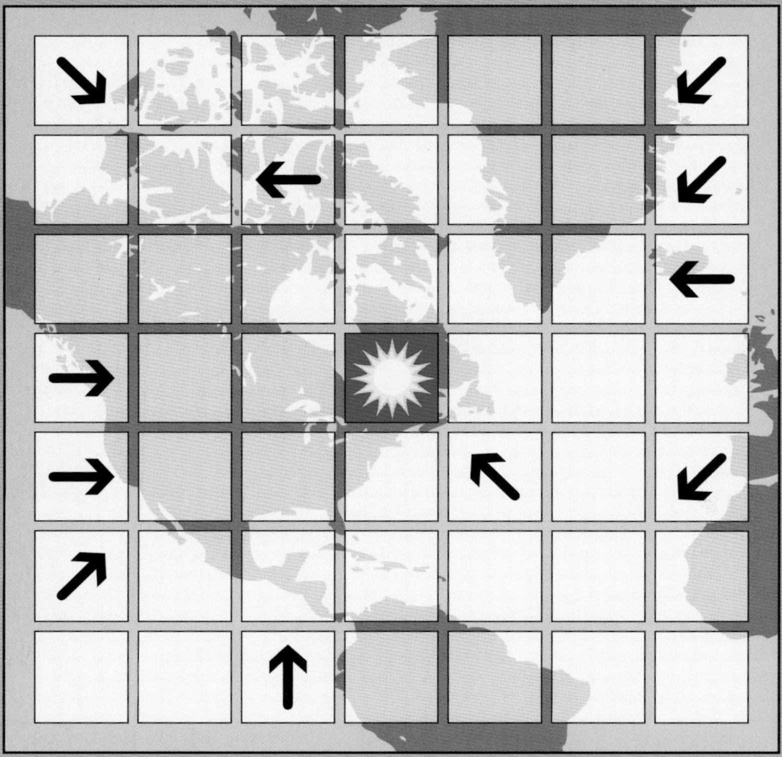

↘						↙
		←				↙
						←
→			☀			
→				↖		↙
↗						
		↑				

TEST ZONE

Find eleven sunny areas.

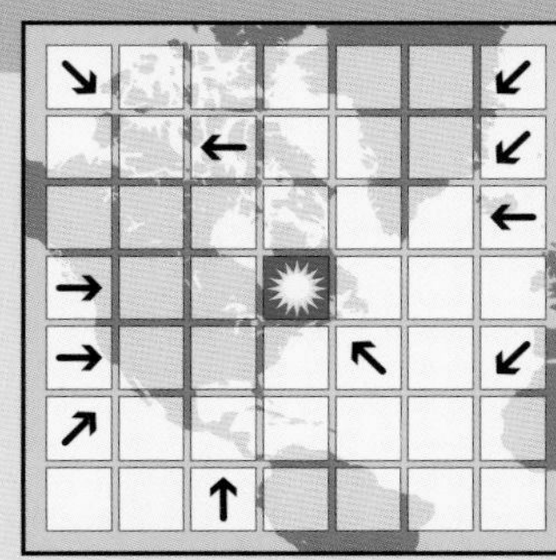

Grid Puzzles

168 SUNNY WEATHER FORECAST

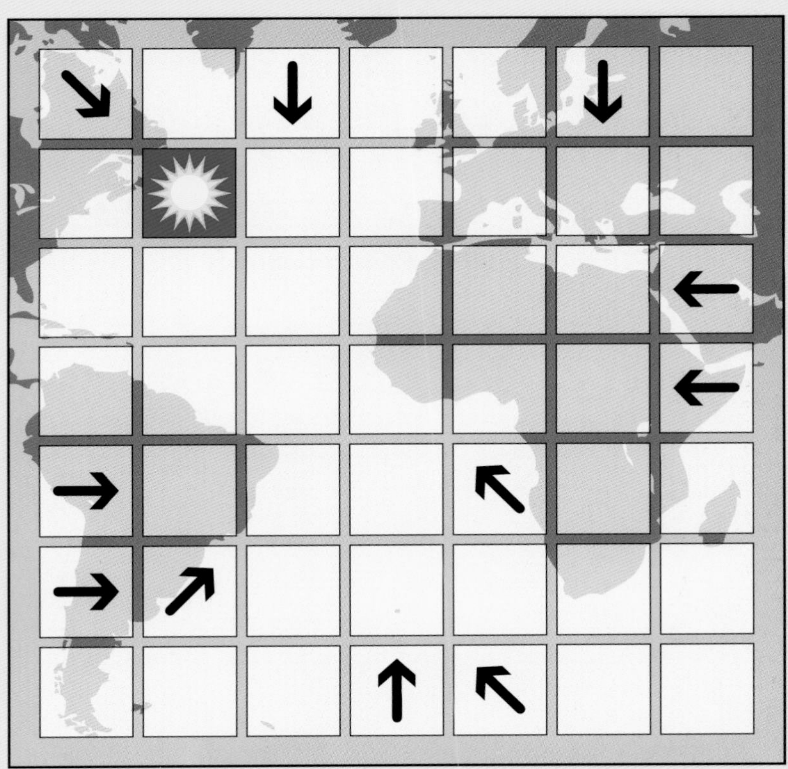

TEST ZONE

Find eleven sunny areas.

Grid Puzzles

DIRECTIONS CONTINUOUS LINE

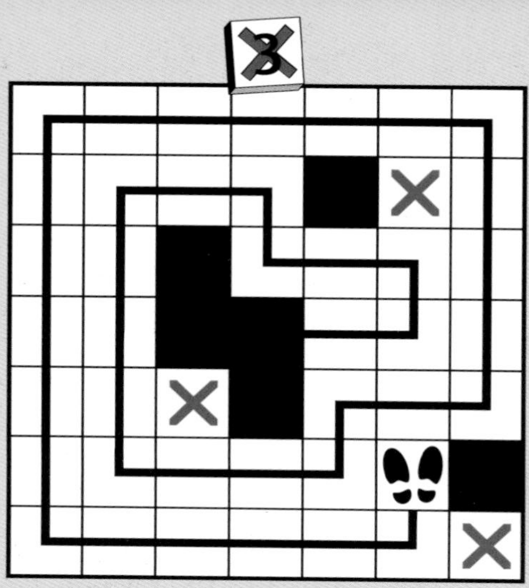

1 Start on a blank square of your choice and connect as many blank squares as possible with one single continuous line.

2 You can only connect squares along vertical and horizontal lines, not along diagonal lines. You must continue the connecting line up until the next obstacle, i.e. the rim of the box, a black square, or a square that has already been used.

3 You can change directions at any obstacle you meet.

4 Each square can only be used once. The number of blank squares that will be left unused is marked in the upper square.

5 There can be more than one solution.

Grid Puzzles

169 CONTINUOUS LINE

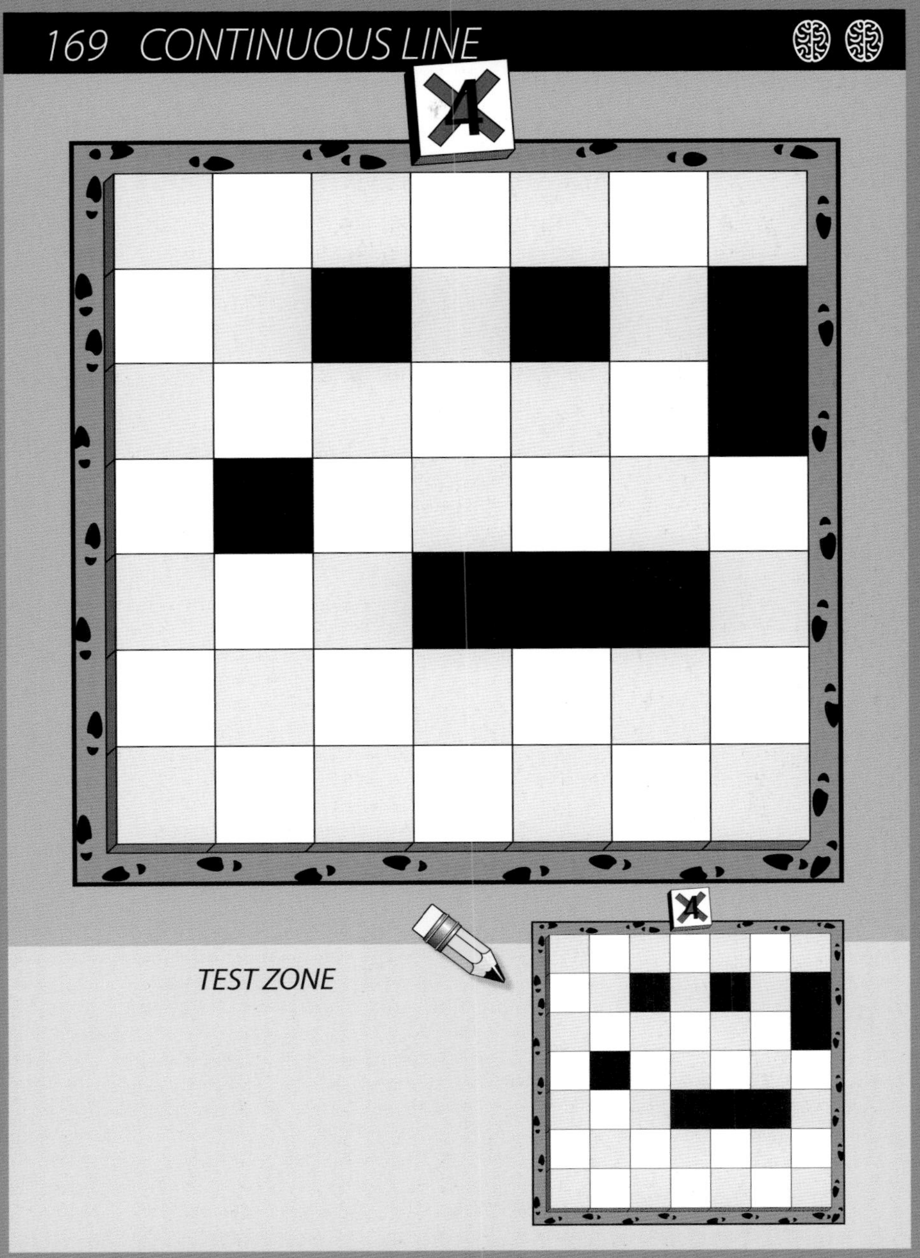

TEST ZONE

Grid Puzzles

170 CONTINUOUS LINE

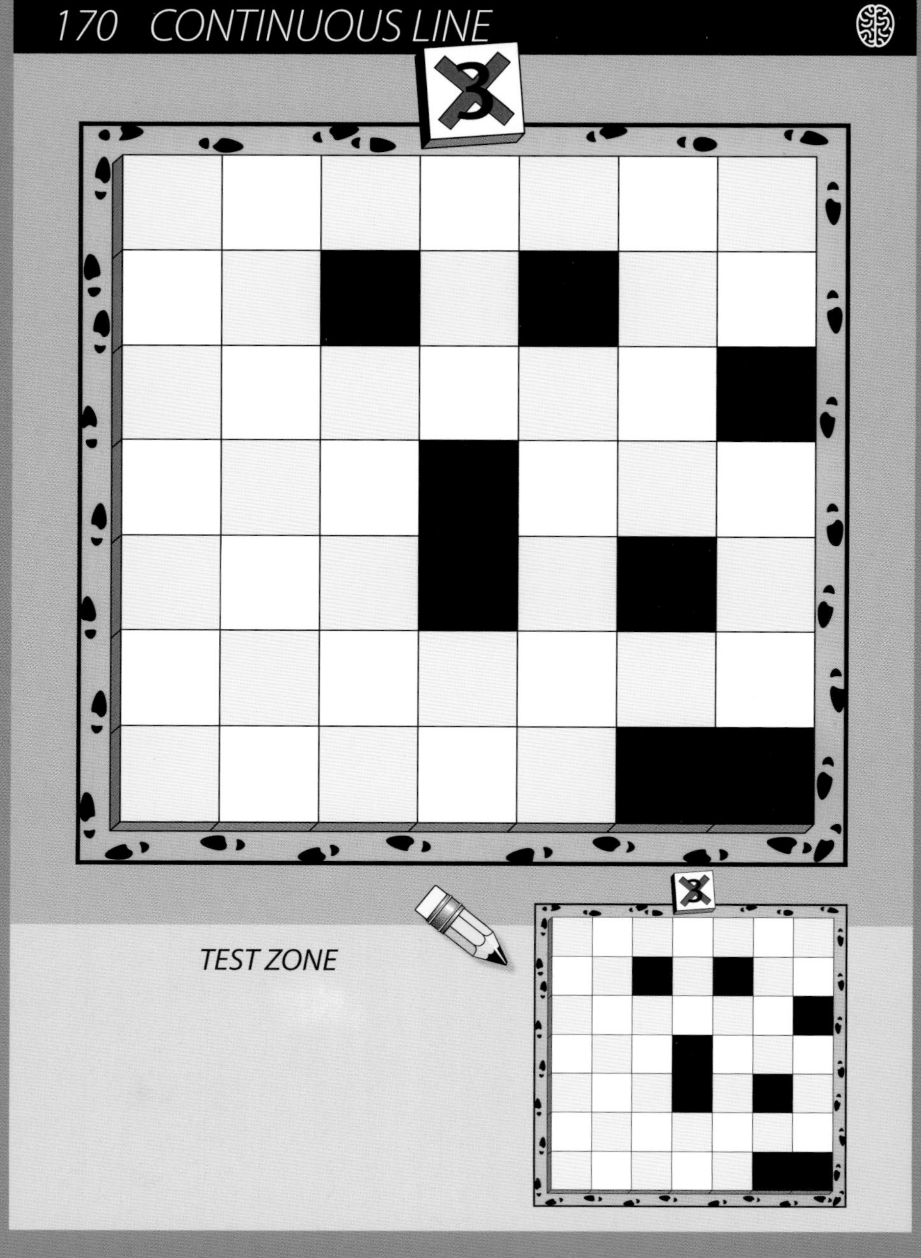

Grid Puzzles

171 CONTINUOUS LINE

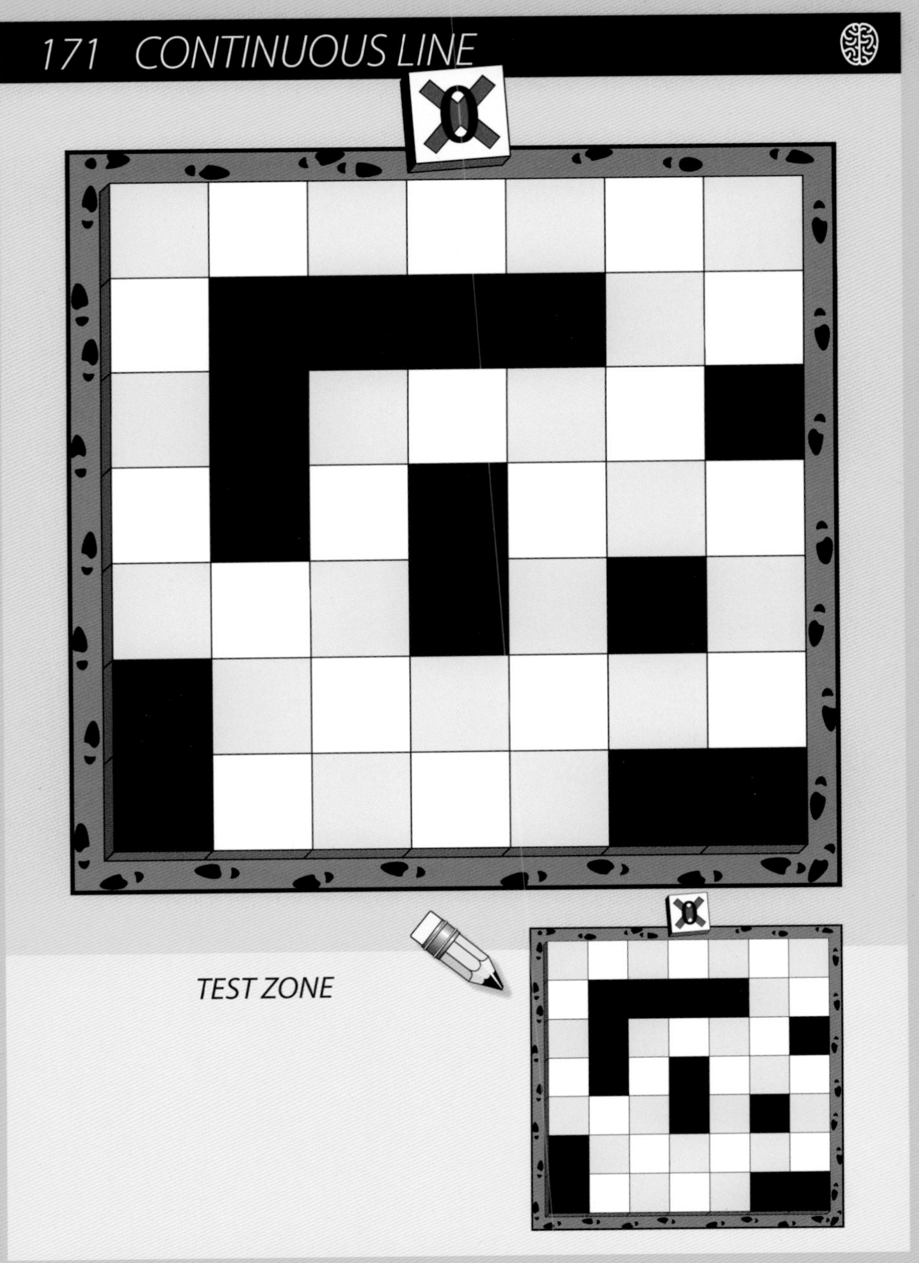

TEST ZONE

Grid Puzzles

172 CONTINUOUS LINE

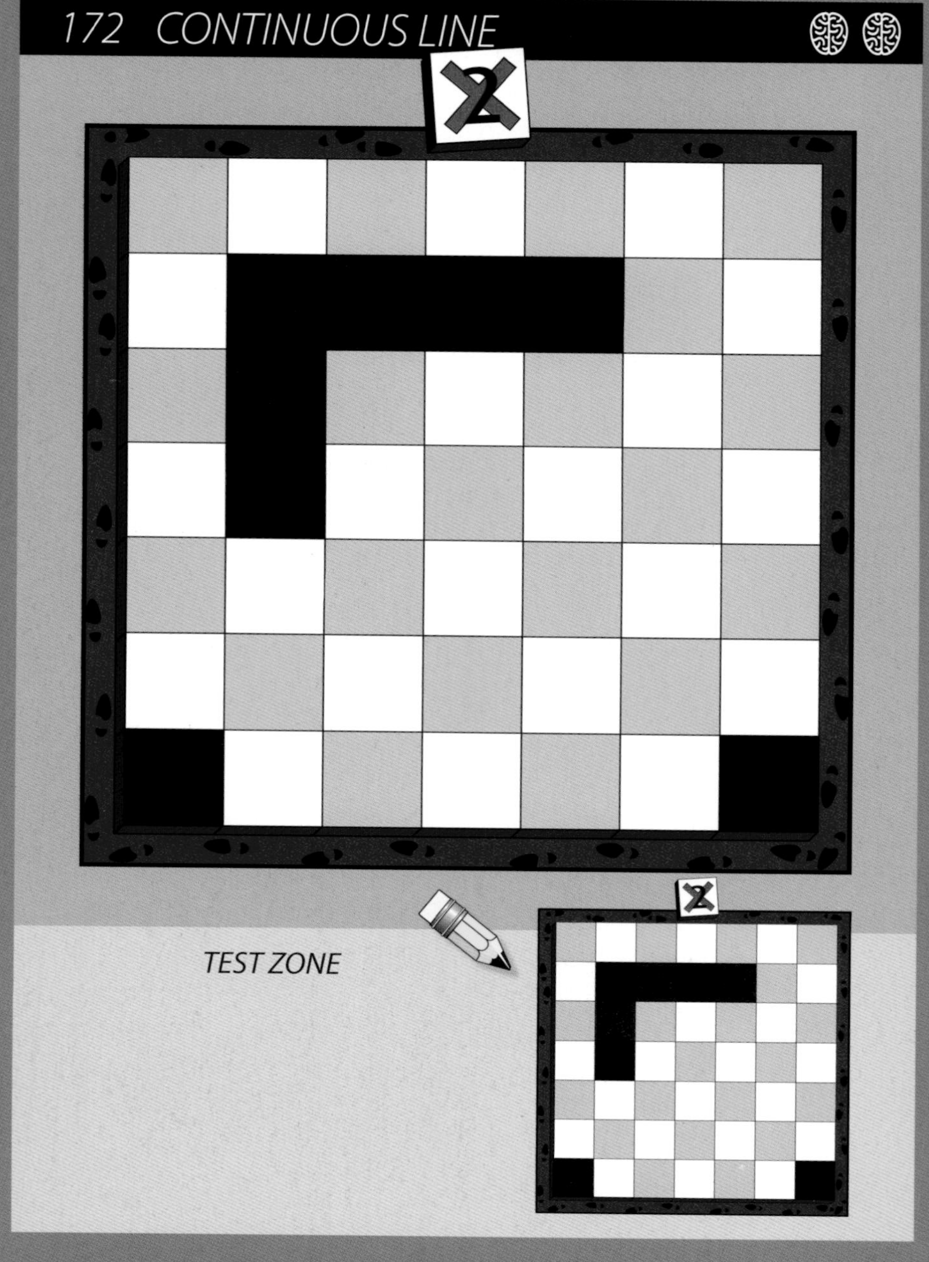

TEST ZONE

Grid Puzzles

173 CONTINUOUS LINE

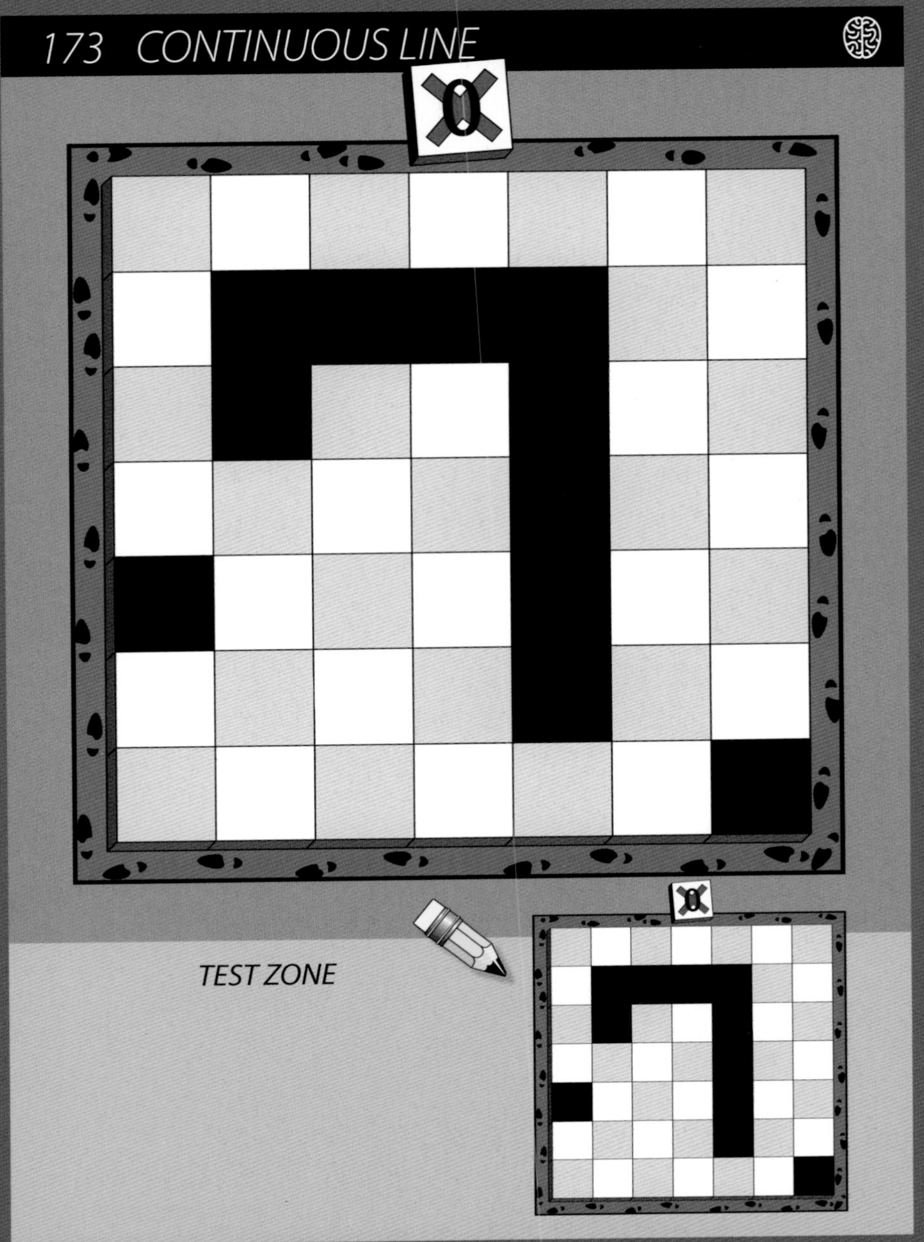

TEST ZONE

Grid Puzzles

174 CONTINUOUS LINE

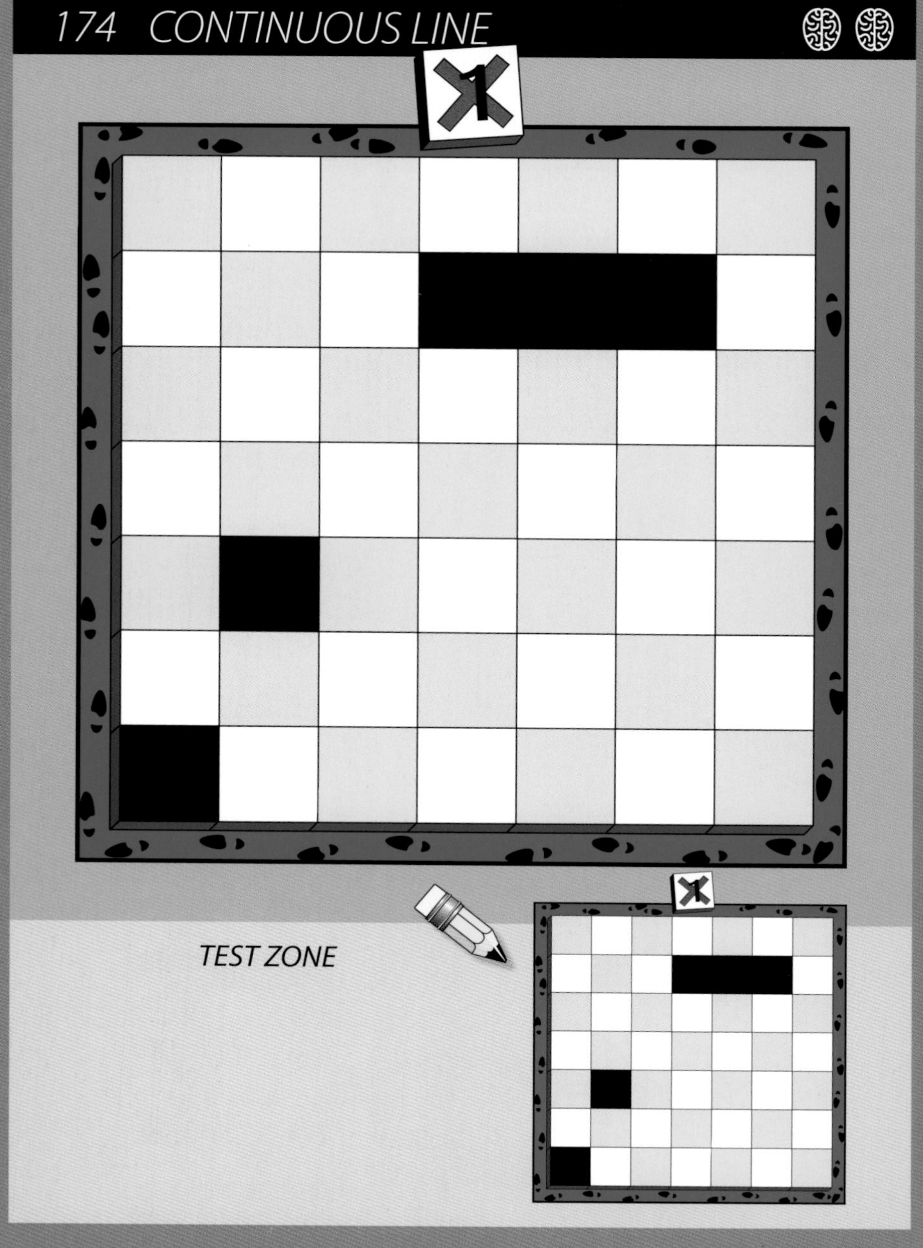

TEST ZONE

Grid Puzzles

175 CONTINUOUS LINE

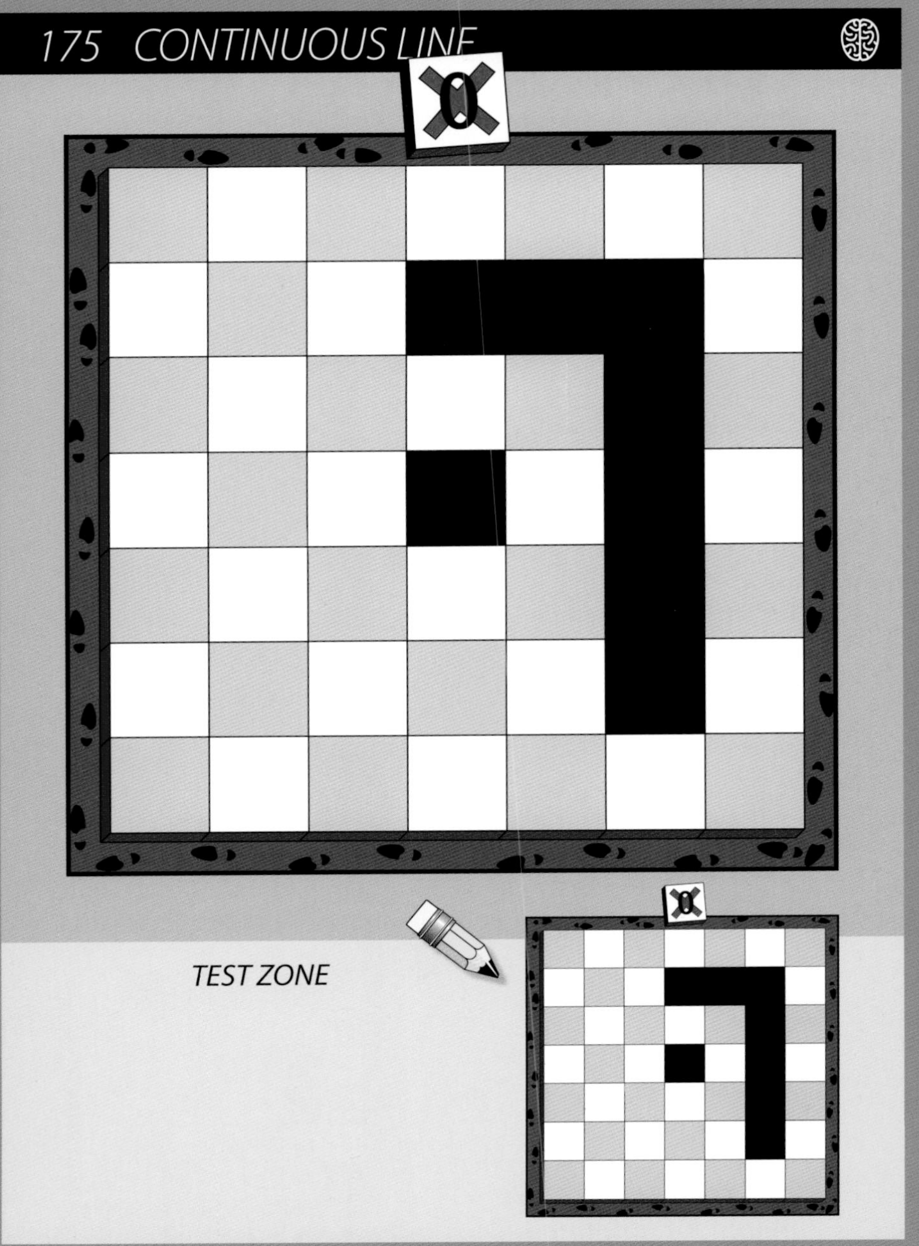

TEST ZONE

Grid Puzzles

176 CONTINUOUS LINE

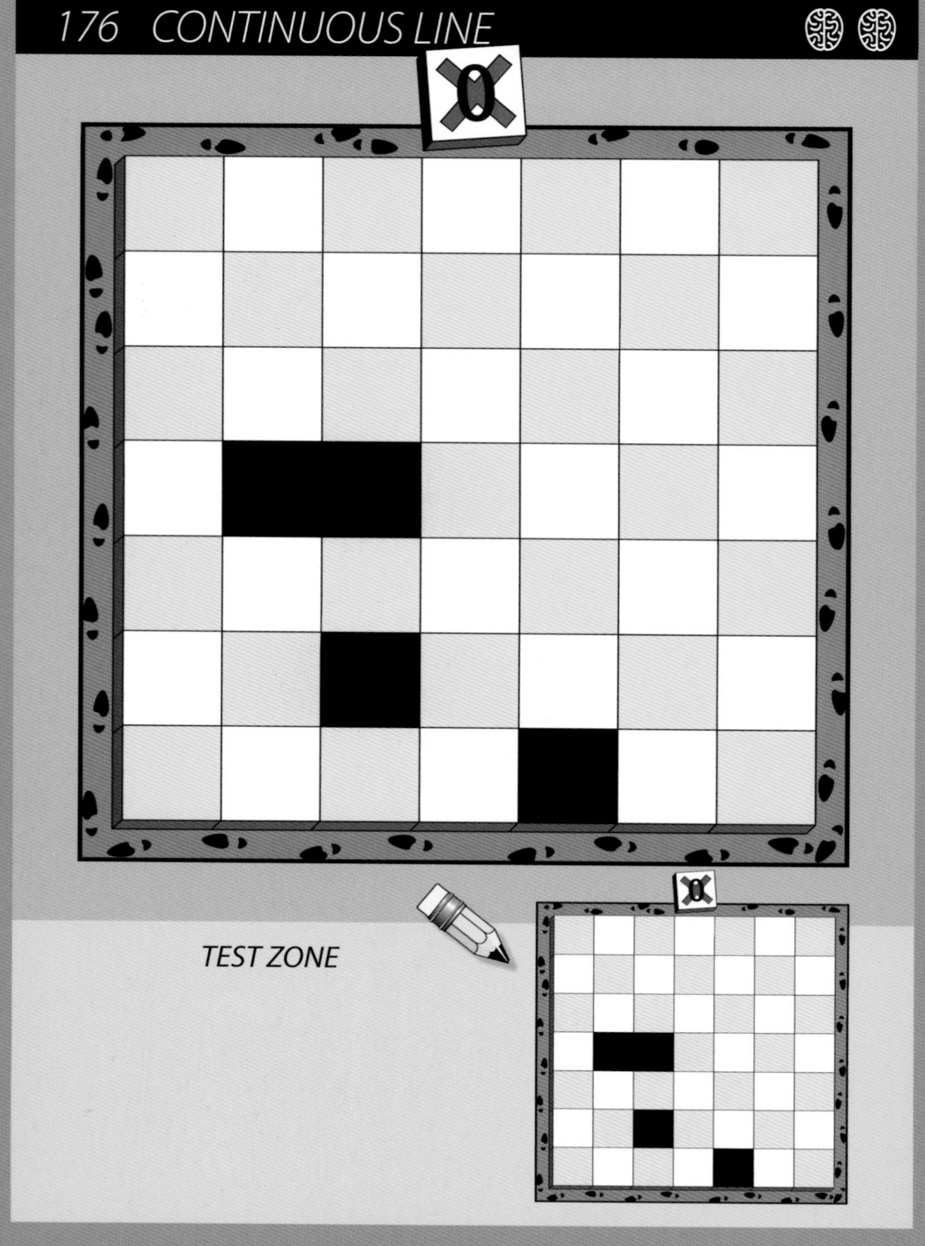

TEST ZONE

Grid Puzzles

177 CONTINUOUS LINE

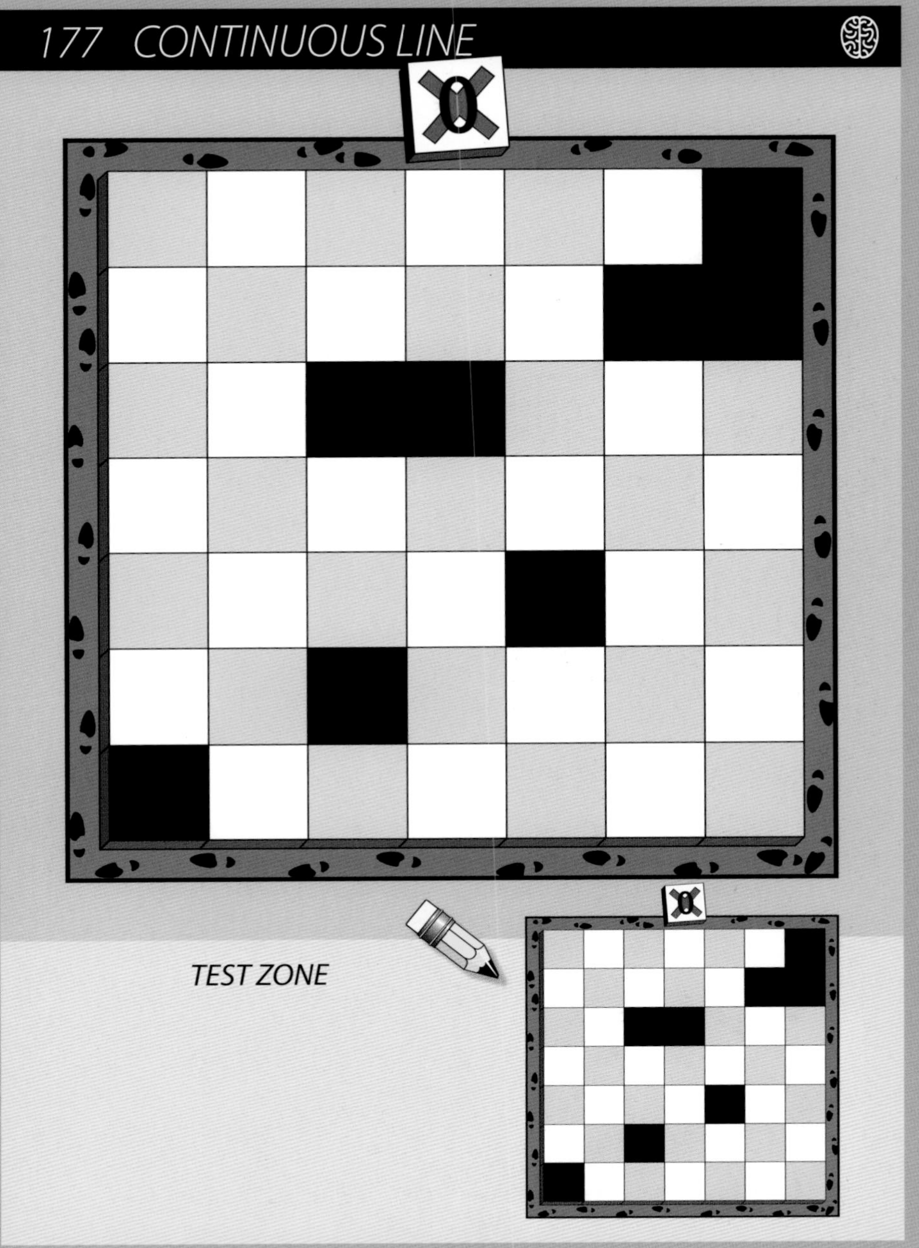

Grid Puzzles

178 CONTINUOUS LINE

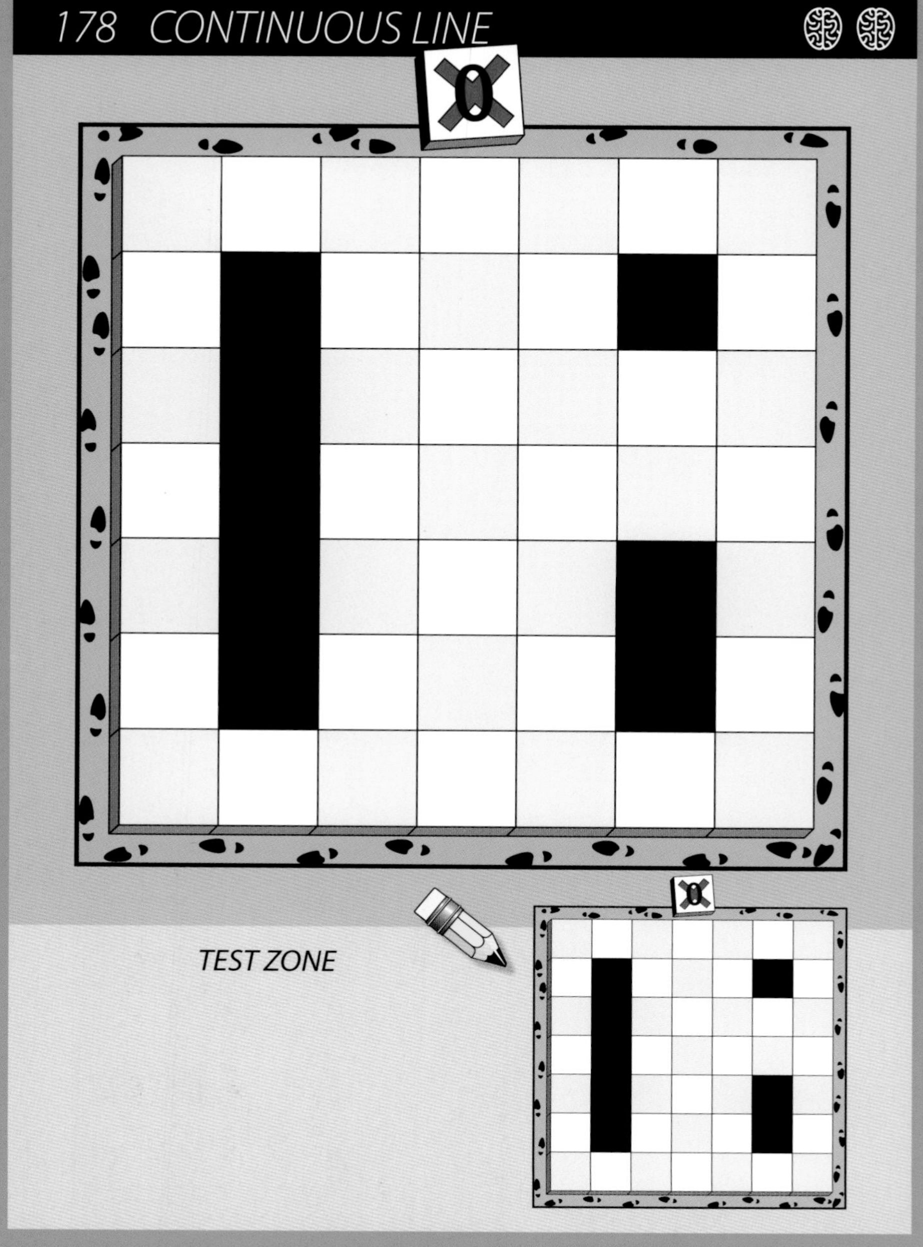

TEST ZONE

Grid Puzzles

179 CONTINUOUS LINE

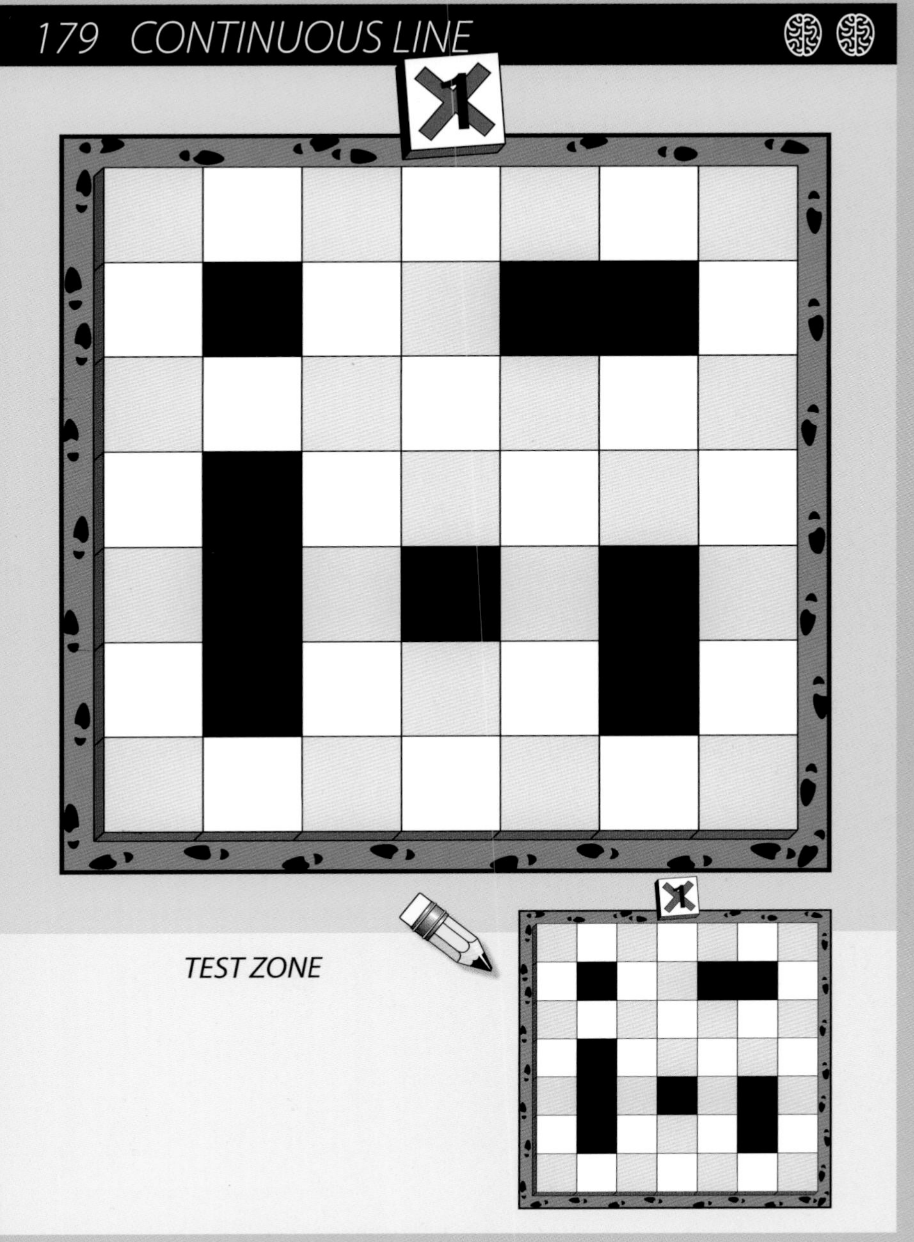

TEST ZONE

Grid Puzzles

DIRECTIONS FOOTBALL & GOLF

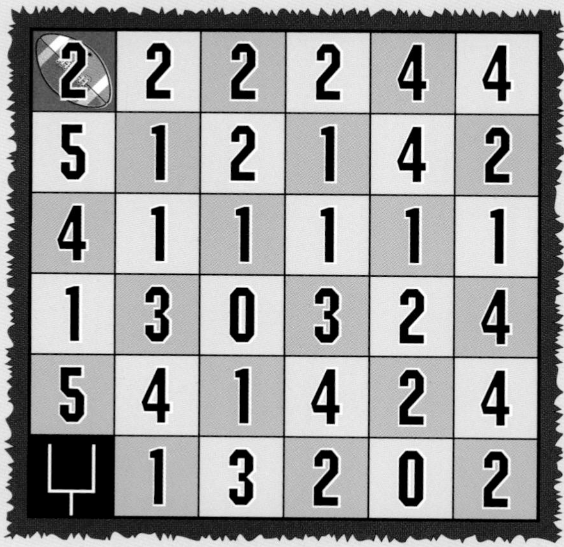

1 *Draw the shortest way from the ball to the goal.*

2 *You can only move along vertical and horizontal lines, not along diagonal lines.*

3 *The figure on each square indicates the number of squares the ball must be moved in the same direction.*

4 *You can change directions at each stop.*

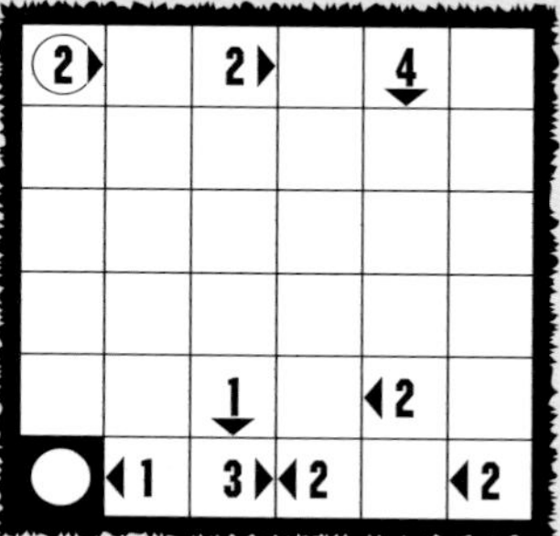

Grid Puzzles

180 GOLF

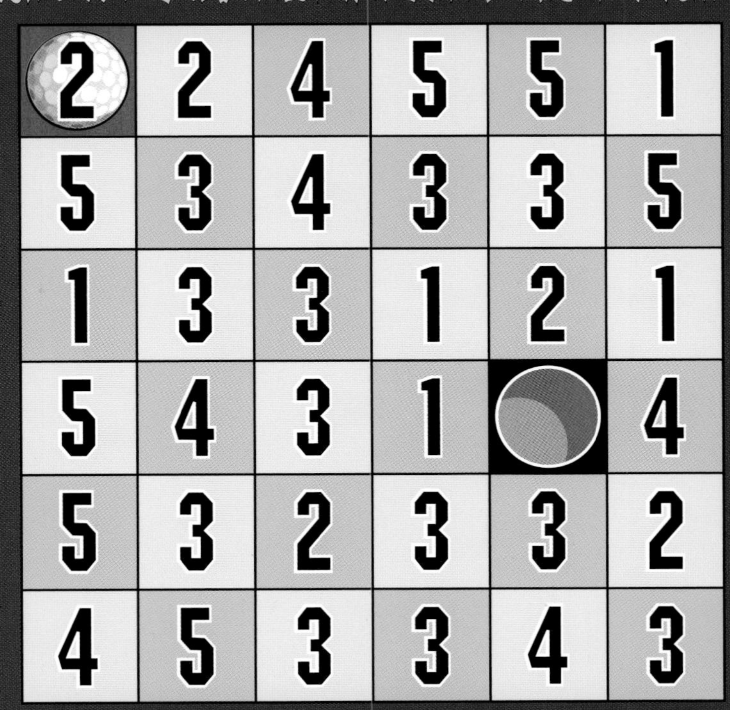

TEST ZONE

2	2	4	5	5	1
5	3	4	3	3	5
1	3	3	1	2	1
5	4	3	1		4
5	3	2	3	3	2
4	5	3	3	4	3

Grid Puzzles

181 FOOTBALL

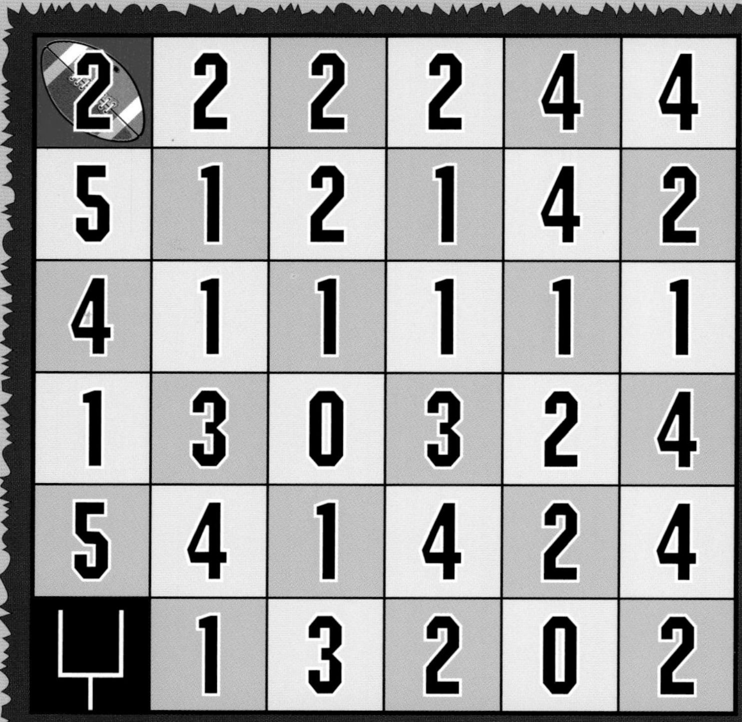

TEST ZONE

2 2 2 2 4 4
5 1 2 1 4 2
4 1 1 1 1 1
1 3 0 3 2 4
5 4 1 4 2 4
1 3 2 0 2

Grid Puzzles

182 GOLF

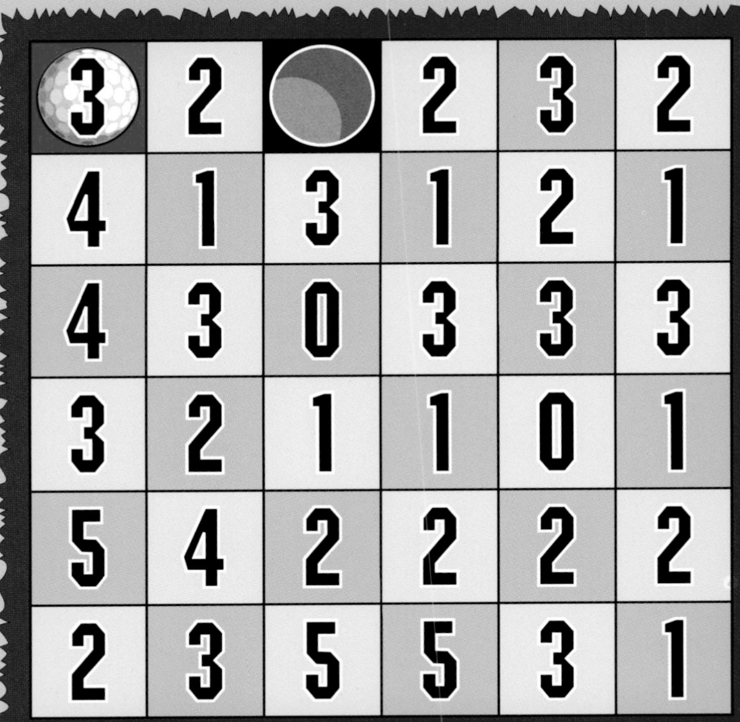

TEST ZONE

3	2		2	3	2
4	1	3	1	2	1
4	3	0	3	3	3
3	2	1	1	0	1
5	4	2	2	2	2
2	3	5	5	3	1

Grid Puzzles

183 FOOTBALL

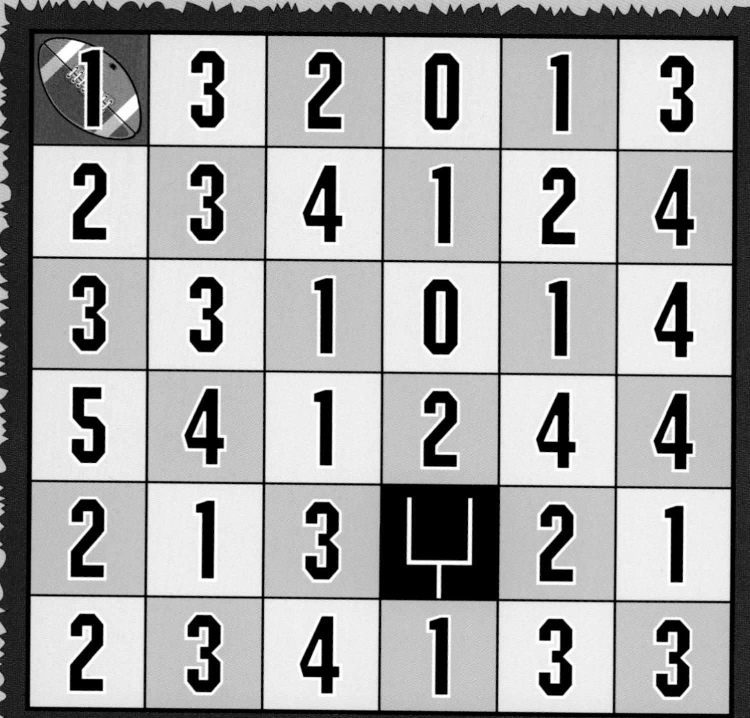

TEST ZONE

1	3	2	0	1	3
2	3	4	1	2	4
3	3	1	0	1	4
5	4	1	2	4	4
2	1	3		2	1
2	3	4	1	3	3

Grid Puzzles

184 GOLF

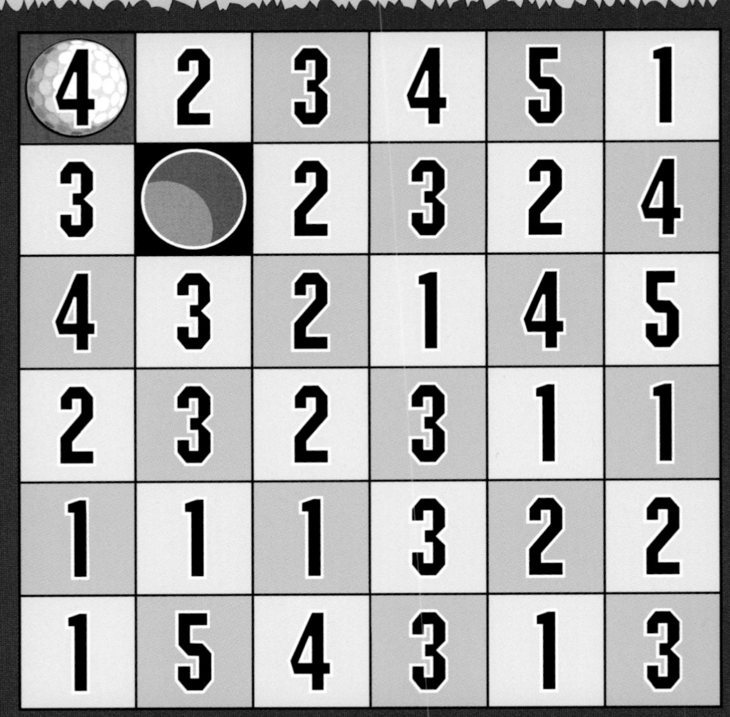

TEST ZONE

4	2	3	4	5	1
3		2	3	2	4
4	3	2	1	4	5
2	3	2	3	1	1
1	1	1	3	2	2
1	5	4	3	1	3

Grid Puzzles

185 FOOTBALL

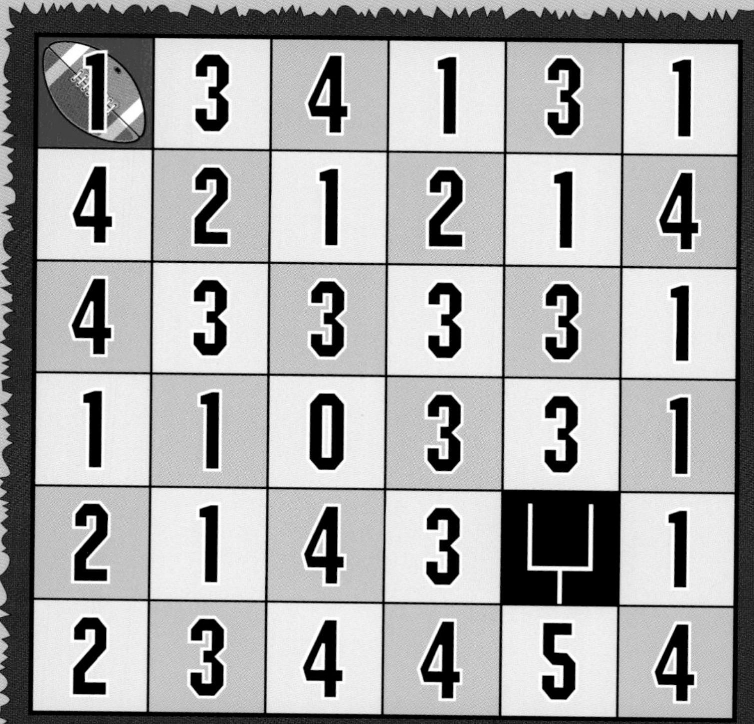

TEST ZONE

1	3	4	1	3	1
4	2	1	2	1	4
4	3	3	3	3	1
1	1	0	3	3	1
2	1	4	3		1
2	3	4	4	5	4

Grid Puzzles

186 GOLF

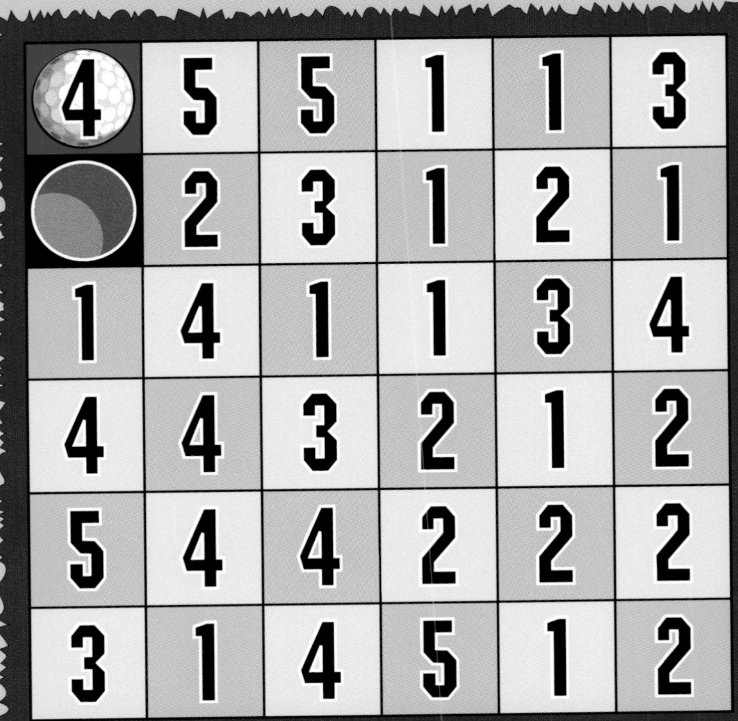

TEST ZONE

4	5	5	1	1	3
	2	3	1	2	1
1	4	1	1	3	4
4	4	3	2	1	2
5	4	4	2	2	2
3	1	4	5	1	2

Grid Puzzles

187 FOOTBALL

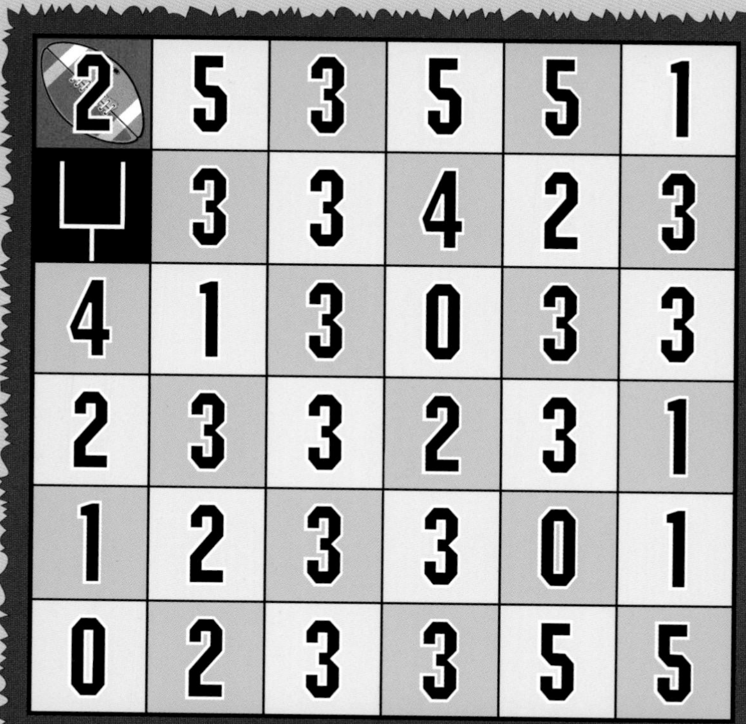

TEST ZONE

2	5	3	5	5	1
	3	3	4	2	3
4	1	3	0	3	3
2	3	3	2	3	1
1	2	3	3	0	1
0	2	3	3	5	5

Grid Puzzles

188 GOLF

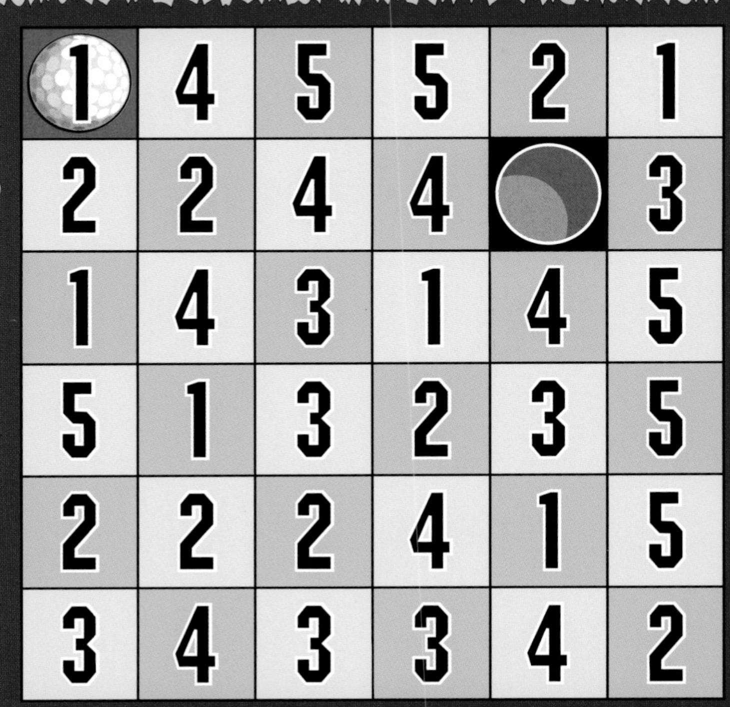

TEST ZONE

1	4	5	5	2	1
2	2	4	4		3
1	4	3	1	4	5
5	1	3	2	3	5
2	2	2	4	1	5
3	4	3	3	4	2

Grid Puzzles

189 FOOTBALL

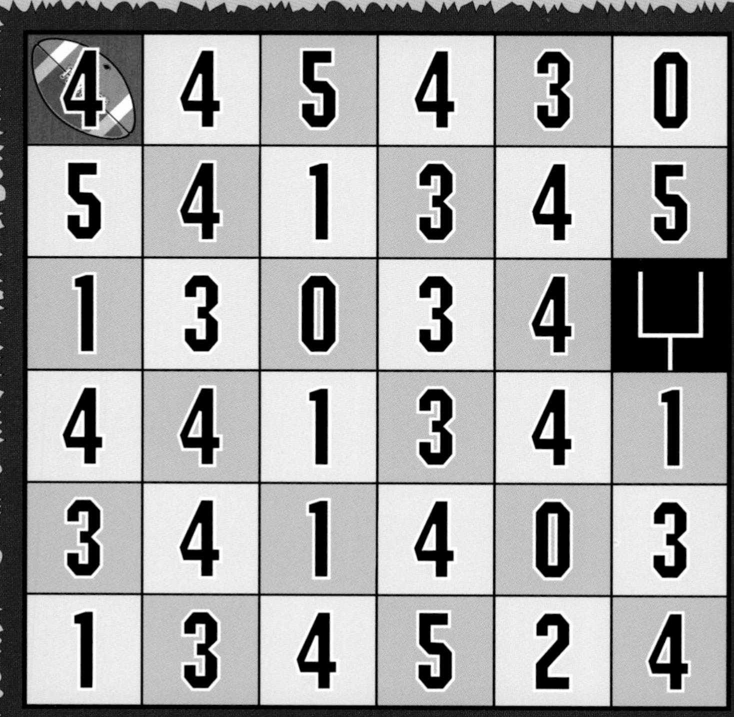

TEST ZONE

4	4	5	4	3	0
5	4	1	3	4	5
1	3	0	3	4	
4	4	1	3	4	1
3	4	1	4	0	3
1	3	4	5	2	4

Grid Puzzles

190 GOLF

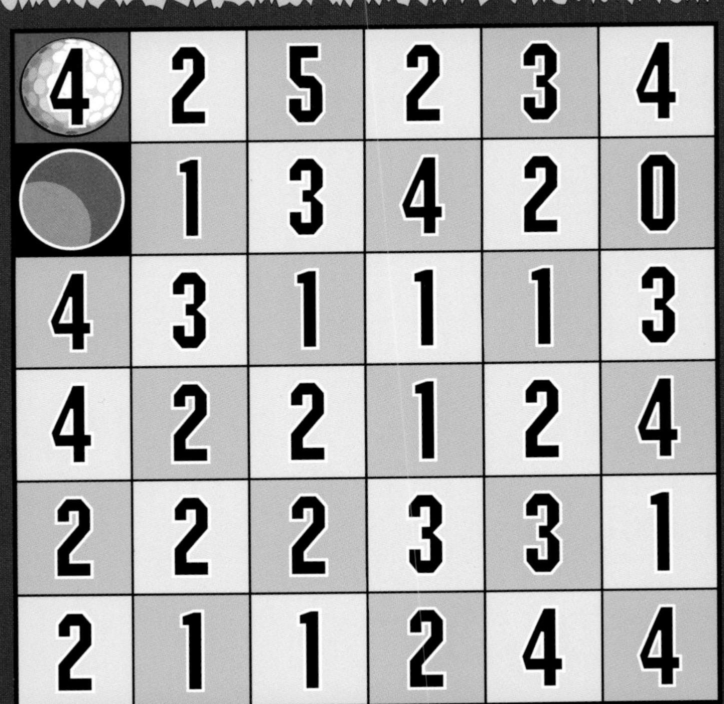

TEST ZONE

4	2	5	2	3	4
	1	3	4	2	0
4	3	1	1	1	3
4	2	2	1	2	4
2	2	2	3	3	1
2	1	1	2	4	4

Grid Puzzles

DIRECTIONS BINARY

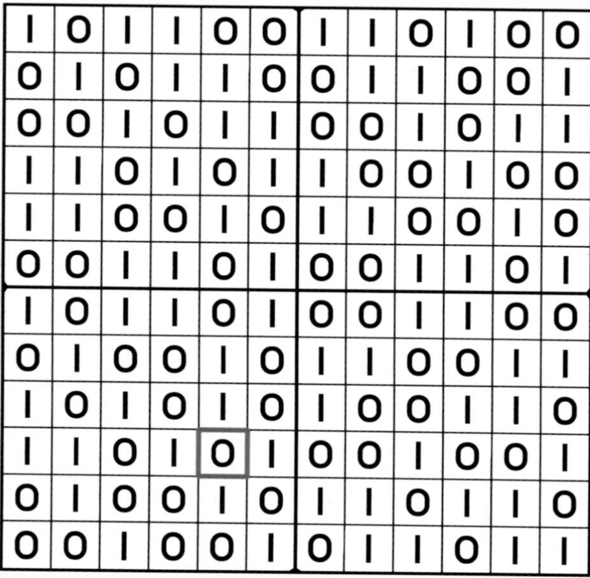

I	O	I	I	O	O	I	I	O	I	O	O
O	I	O	I	I	O	O	I	I	O	O	I
O	O	I	O	I	I	O	O	I	O	I	I
I	I	O	I	O	I	I	O	O	I	O	O
I	I	O	O	I	O	I	I	O	O	I	O
O	O	I	I	O	I	O	O	I	I	O	I
I	O	I	I	O	I	O	O	I	I	O	O
O	I	O	O	I	O	I	I	O	O	I	I
I	O	I	O	I	O	I	O	O	I	I	O
I	I	O	I	O	I	O	O	I	O	O	I
O	I	O	O	I	O	I	I	O	I	I	O
O	O	I	O	O	I	O	I	I	O	I	I

1 *Complete the grid with zeros and ones until there are just as many zeros and ones in every row and every column.*

2 *No more than two of the same number can be next to or under each other.*

3 *Rows or columns with exactly the same content are not allowed.*

4 *If the assertion is true, write a O in the red square if false a I.*

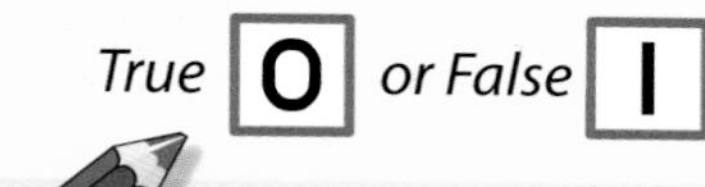

Grid Puzzles

191 BINARY 12 x 12

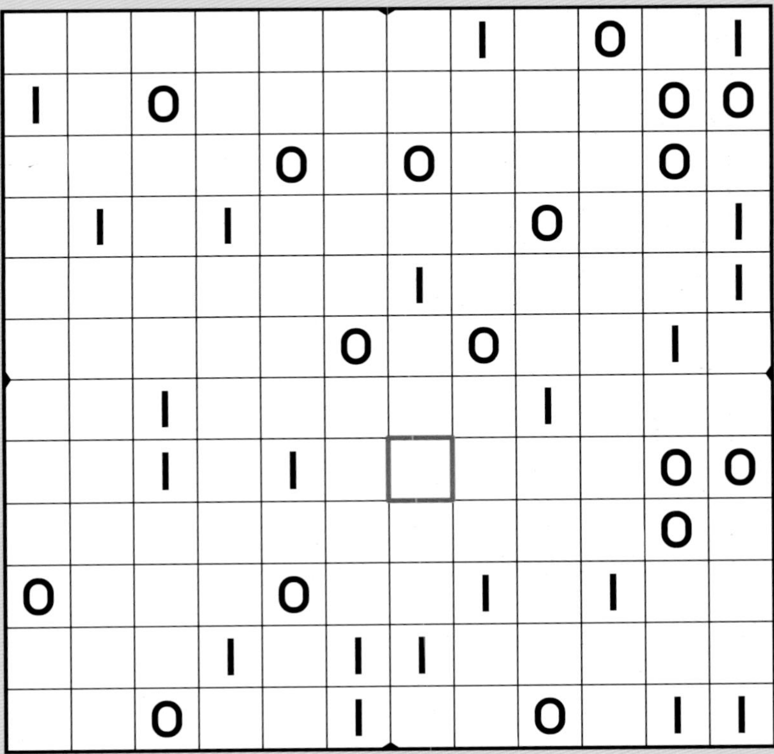

							1		0		1
1		0								0	0
				0		0				0	
	1		1					0			1
						1					1
					0		0			1	
		1						1			
		1		1						0	0
										0	
0				0			1		1		
			1		1	1					
		0			1			0		1	1

Wrestling was originally a Japanese sport.

True 0 *or False* 1

Grid Puzzles

192 BINARY 12 x 12

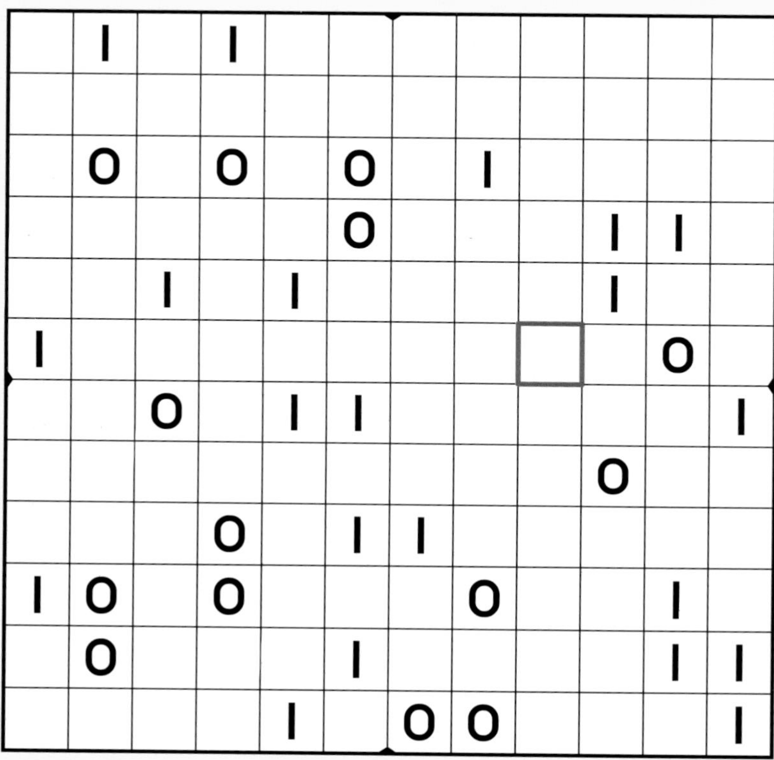

Bakelite is a rock.

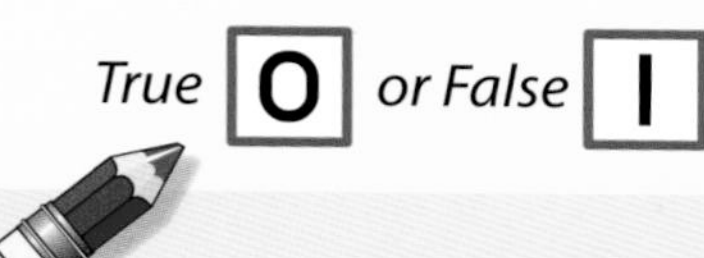

Grid Puzzles

193 BINARY 12 x 12

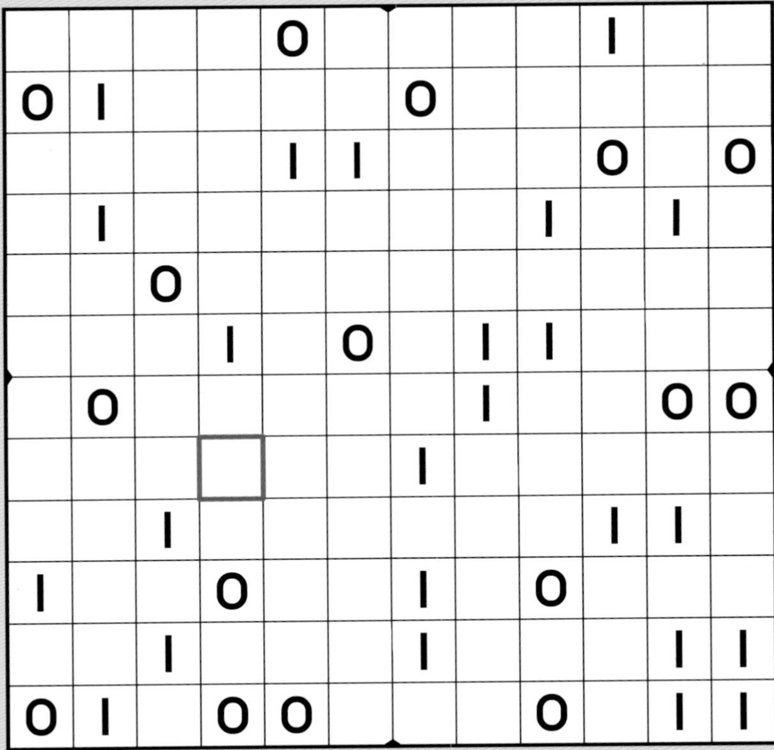

The Amazon is the longest river in the world.

True O *or False* I

Grid Puzzles

194 BINARY 12 x 12

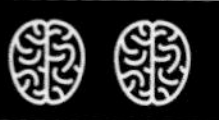

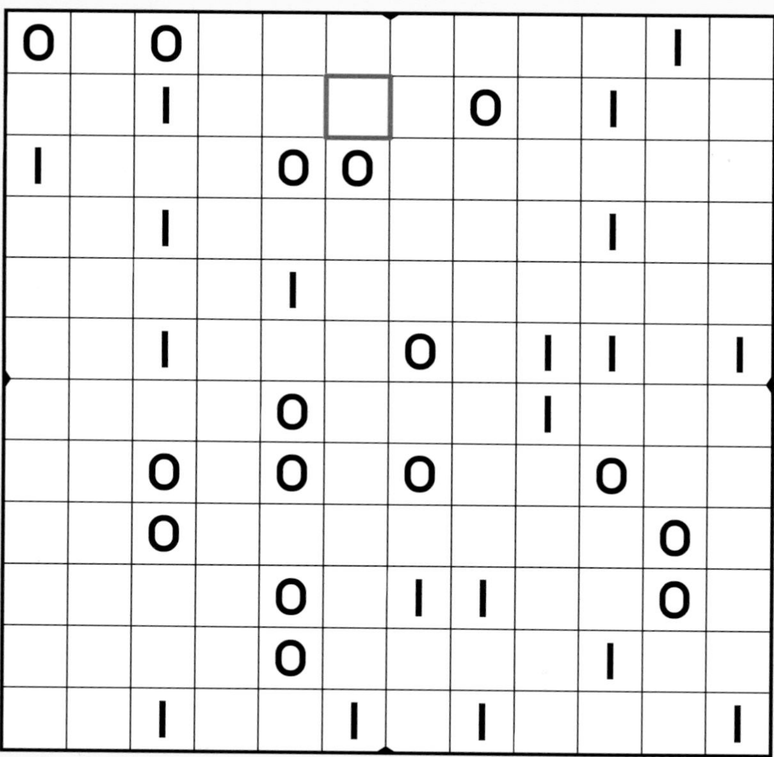

Giraffes sleep standing up and not more than twenty minutes per day.

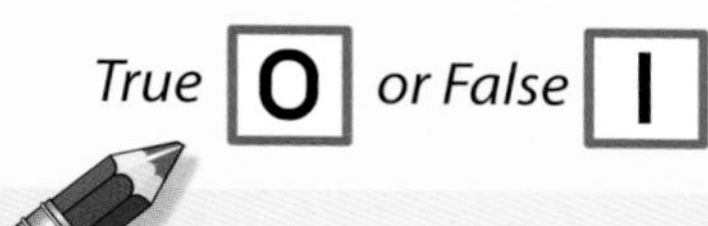

Grid Puzzles

195 BINARY 12 x 12

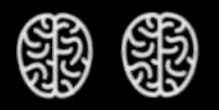

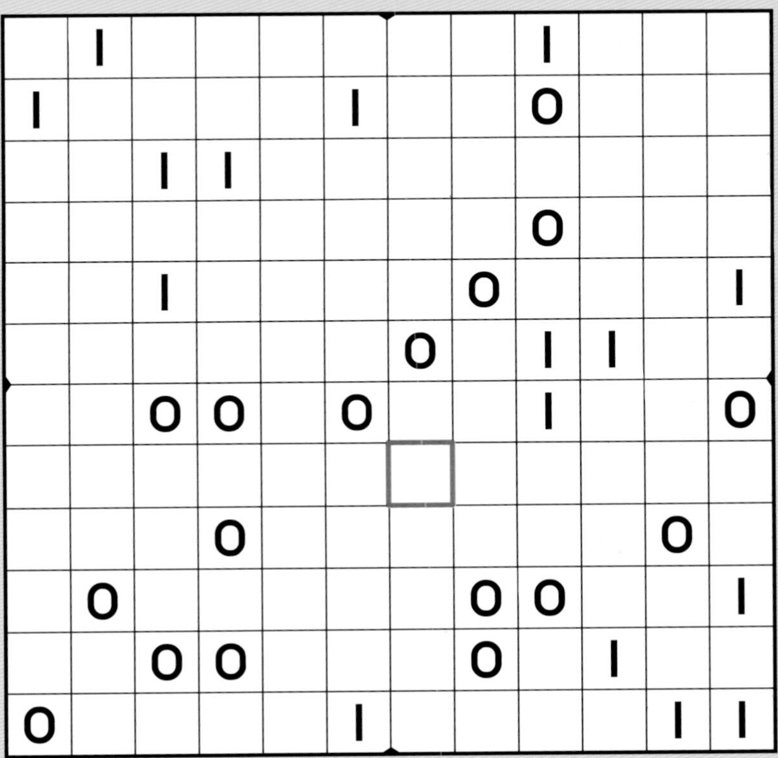

The ballpoint pen was invented during WWII.

True O *or False* I

Grid Puzzles

196 BINARY 12 x 12

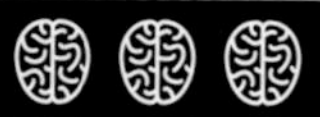

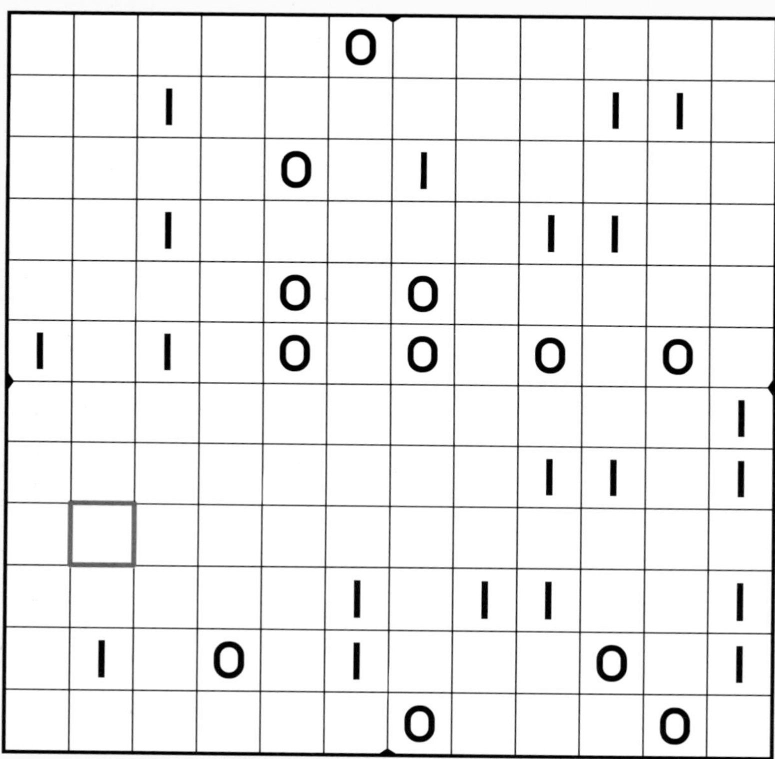

Yellow appears to increase the size of a shape, probably because more cones in the eye are sensitive to yellow than to other colors.

Grid Puzzles

197 BINARY 12 x 12

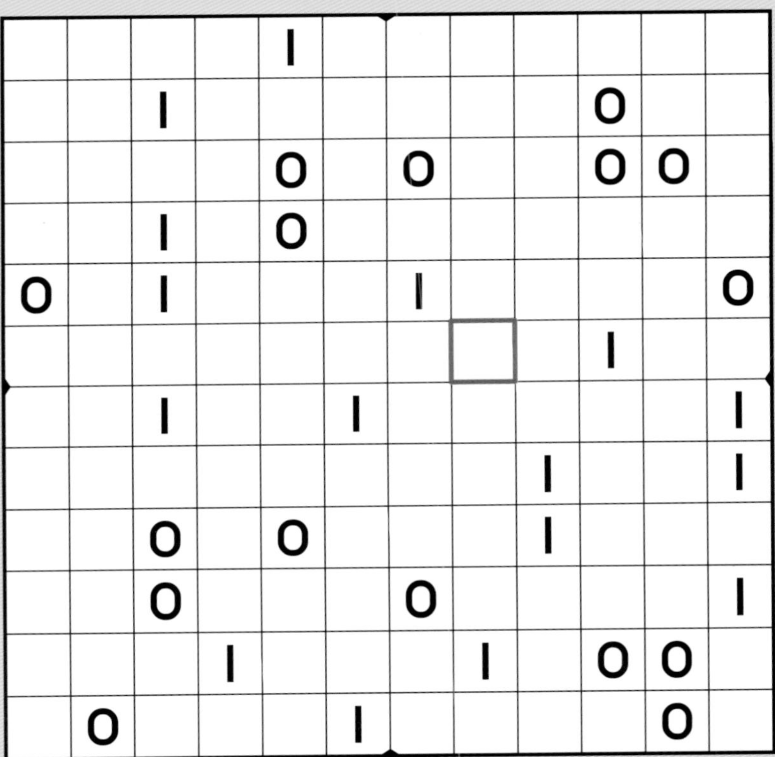

Jeans comes from the French name "blue de Gênes," denim trousers that were shipped in bulk from the port of Genoa in Italy during the Renaissance.

True O *or False* I

Grid Puzzles

198 BINARY 12 x 12

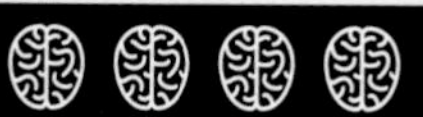

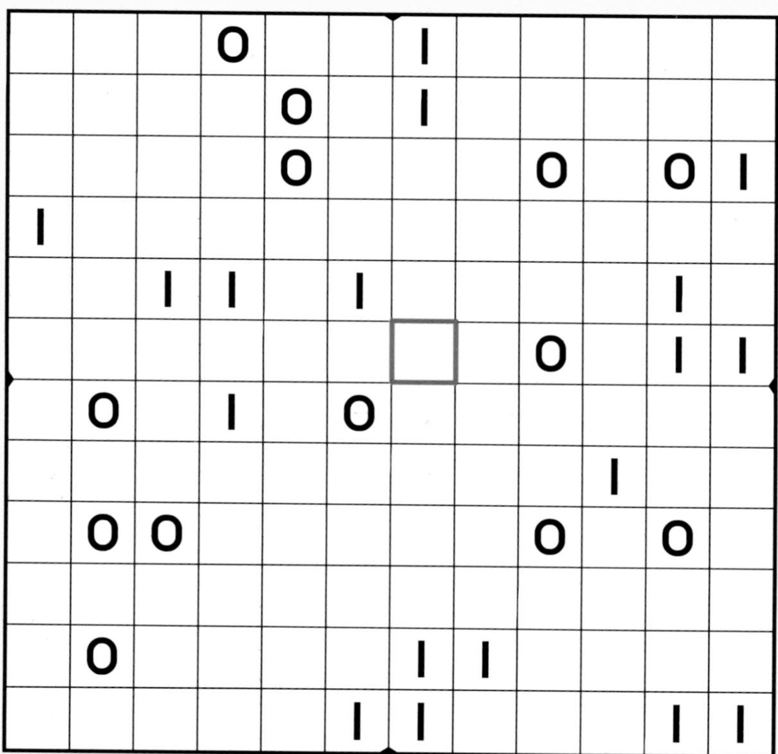

Leonardo da Vinci, painter of the Mona Lisa that hangs in the Louvre in Paris, wrote in reverse from right to left.

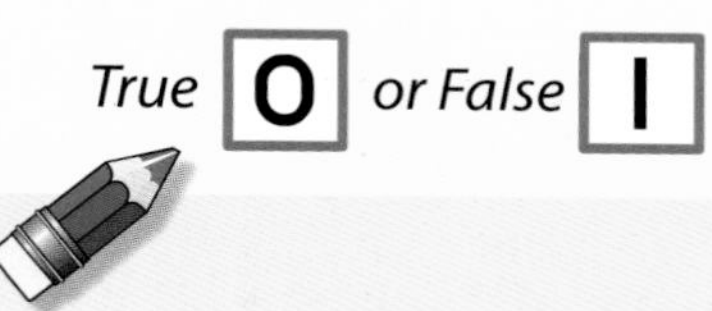

Grid Puzzles

199 BINARY 12 x 12

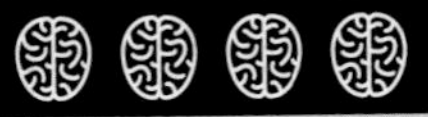

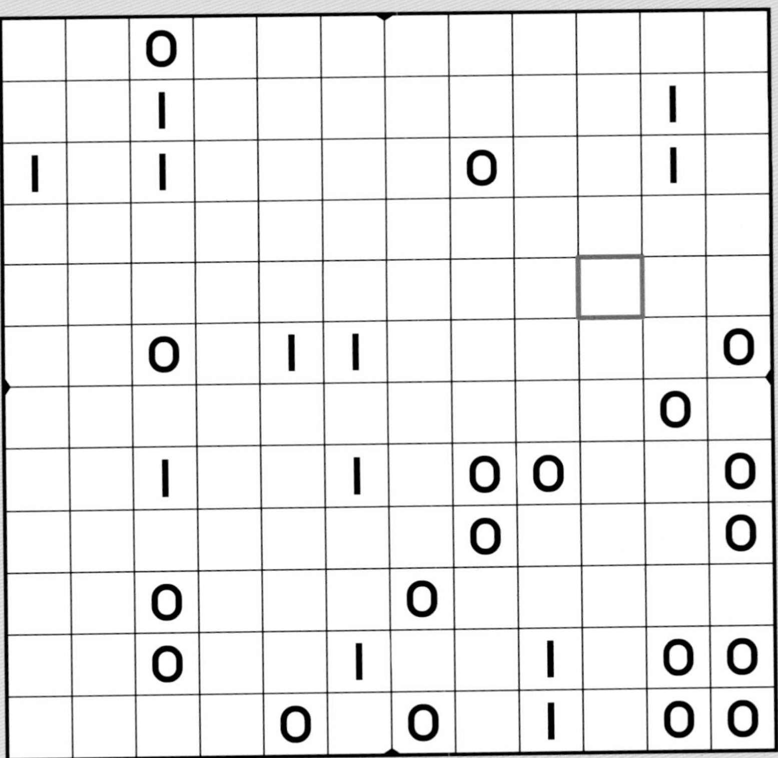

Most land snails are not blind or deaf, but they are hermaphrodites.

True 0 *or False* 1

Grid Puzzles

200 BINARY 14 x 14

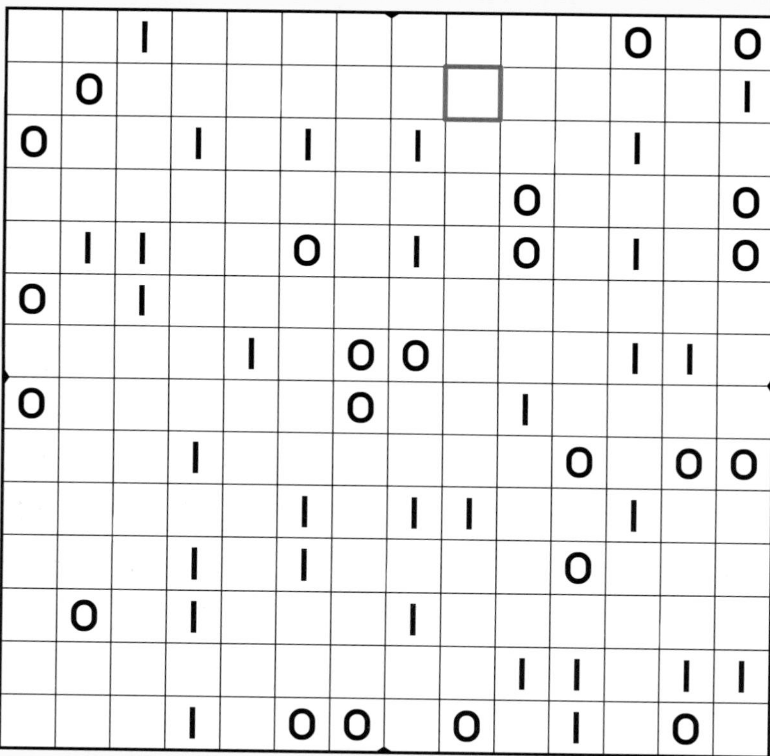

The number of heartbeats per minute of a healthy adult whale is approximately half that of a healthy adult man.

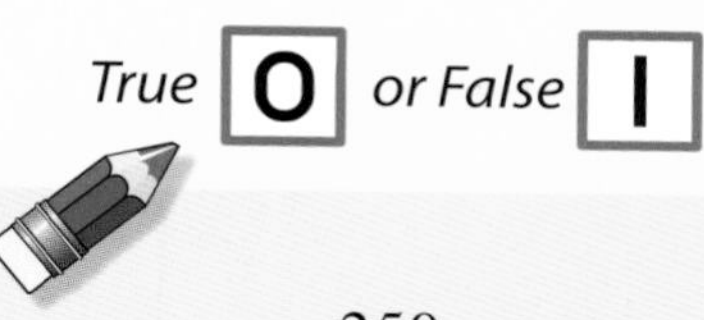

Grid Puzzles

201 BINARY 14 x 14 

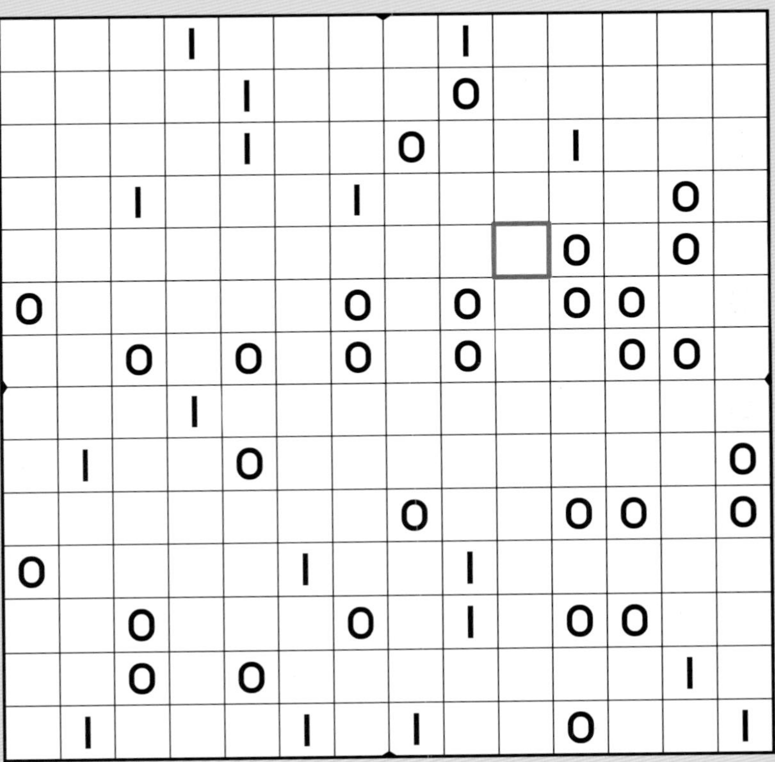

Tora Bora is the highest peak on Bora Bora, an atoll located in the Pacific Ocean between Australia and South America.

True **O** *or False* **I**

Grid Puzzles

202 BINARY 14 x 14

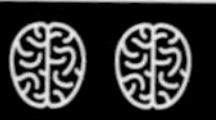

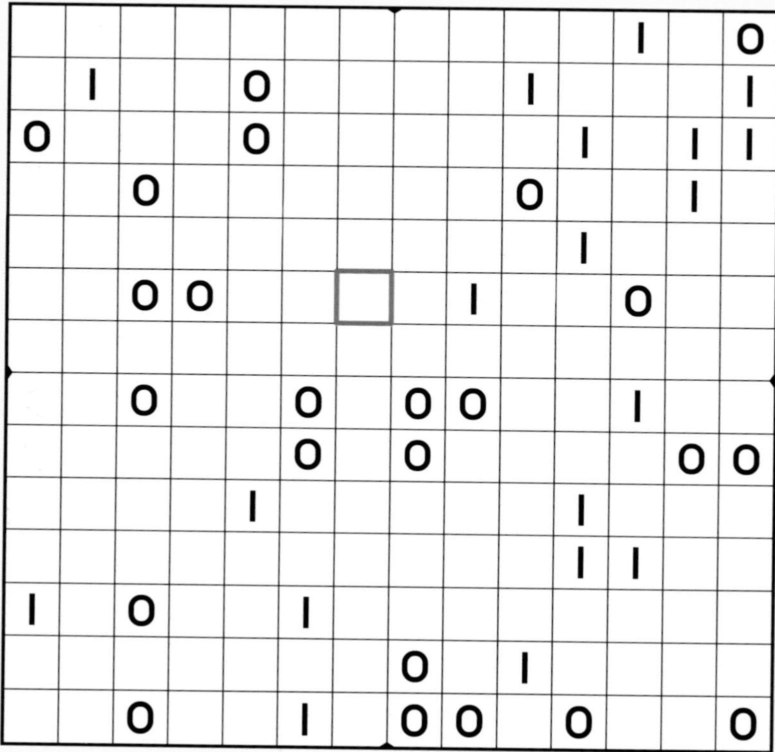

There are no wild tigers in America but there are wild tigers in Africa and Asia.

True O *or False* I

Grid Puzzles

203 BINARY 14 x 14

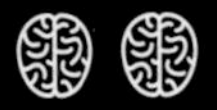

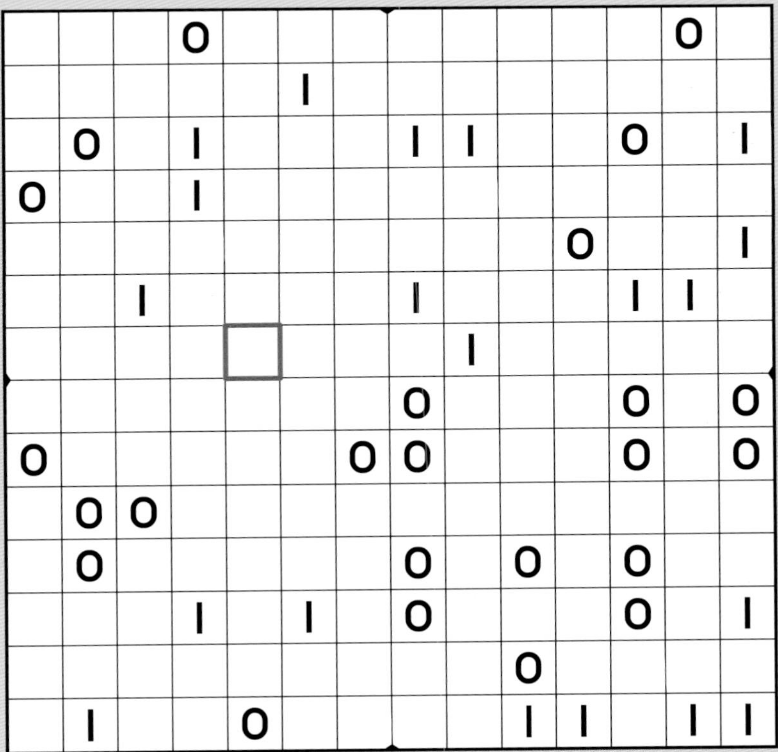

Roller coasters are classified by the U.S. Patent Office as scenic railways.

Grid Puzzles

204 BINARY 14 x 14

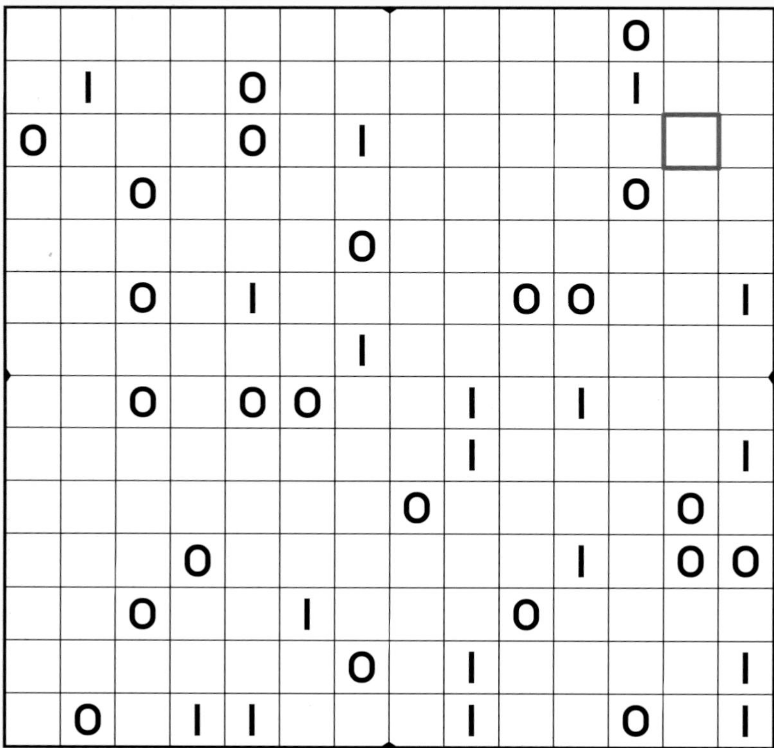

We have five taste sensations:
sweet, sour, salty, bitter, and umami.

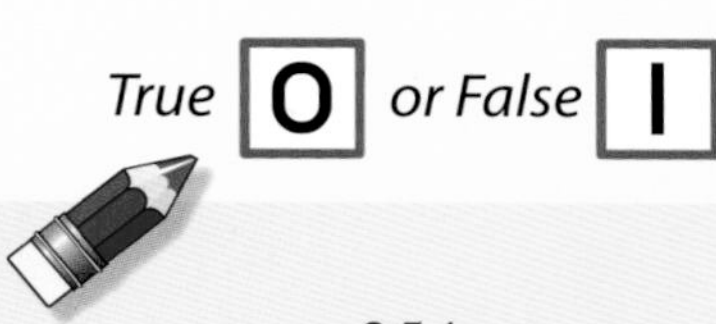

Grid Puzzles

205 BINARY 14 x 14

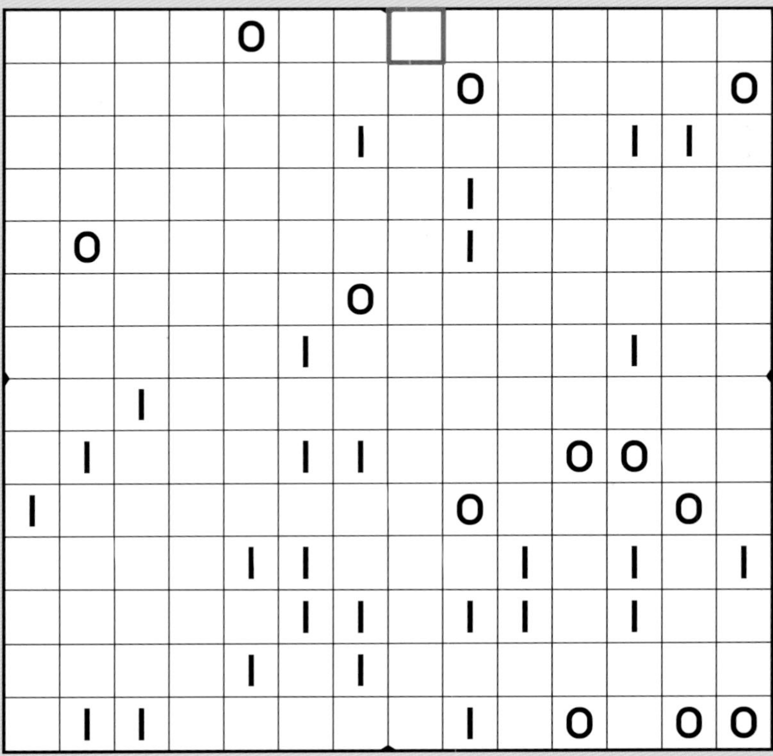

One brain cell can make many thousand connections with other cells. Connections in your brain are also cut, up to twenty billion per day between infancy and puberty.

Grid Puzzles

206 BINARY 14 x 14

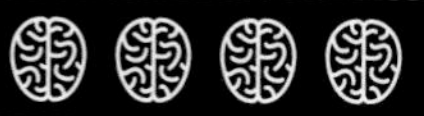

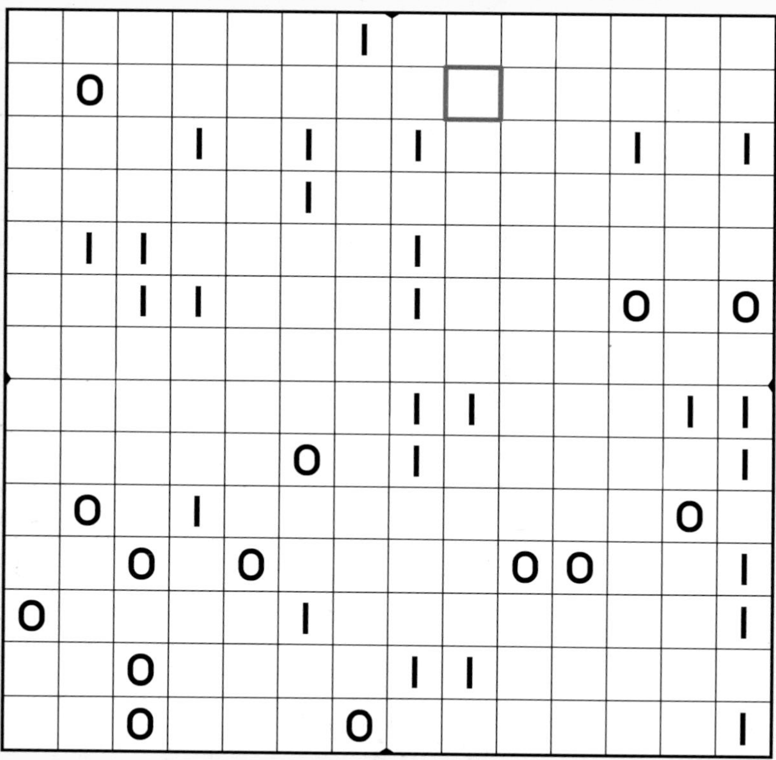

In Rome, the soldier's pay (salarium) was originally salt and the word salary derives from it.

Grid Puzzles

207 BINARY 14 x 14

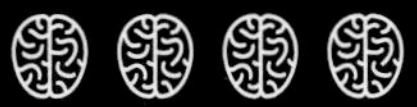

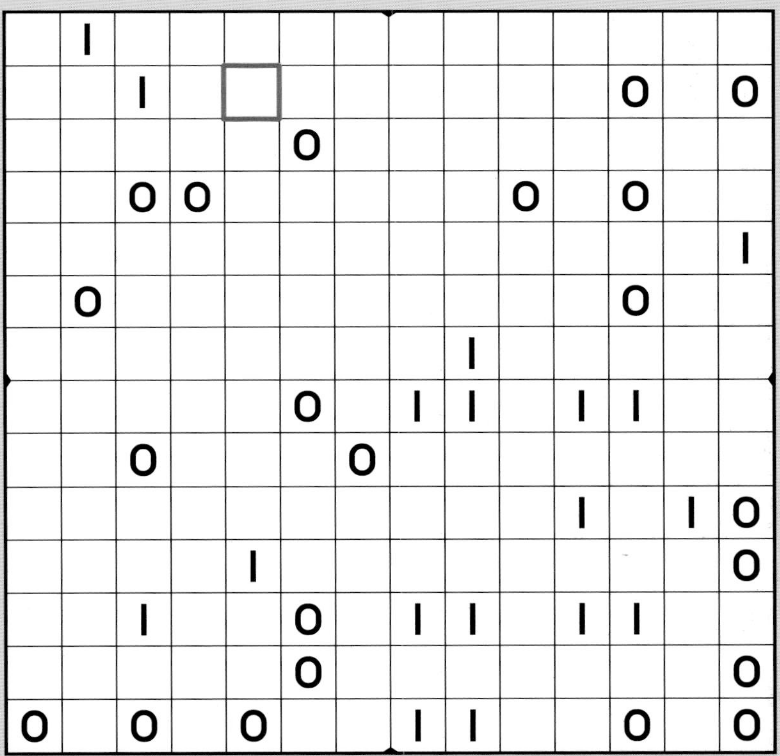

Seals have no ear flaps, but sea lions do.

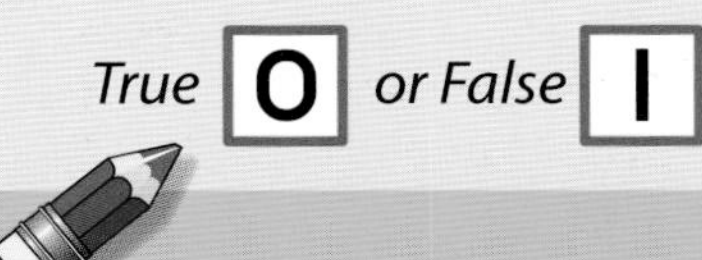

Solutions

1 *Six cakes*
Only the pieces in the upper left and lower right are identical.

2 *Six berries*
The next branch always contains the same number of unripe berries as there were ripe berries on the previous branch.

3 *One gallon of paint*
To paint one of the twenty-four surfaces, you need 1/3 (8/24) of a gallon.
3 surfaces x 1/3 of a gallon=1 gallon

4 *A*
The second and second to last letter of the previous port are the first and last letter of the next port, respectively.

5 *1C2D3A4E5F6B*

6 *Fish E*
All the fish that are swimming in the other direction have scales, just like those on fish E.

7 *Marble 3*
All non-purple marbles are in groups of five and form an angle.

8 *Stack C*
The left and right sides swap, and the order of the dice reverse from top to bottom.

9 *Color H*
The colors are mirrored after the tenth slat.

10 *7*
The following numbers are on all tea bags twice:
19, 27, 45, and 93.

Solutions

11 *Symbol 5*
Each symbol continues building on the previous one. The extra horizontal line in symbol 5 should slant up in order to fit in with symbols 6, 7, and 8.

12 *Eight years*
The number of letters in the two names indicates the number of years the couple has been in love.

13 *Nest S*
Each row has one additional egg.

14 *Set 4*
In all the other sets, the flowers next to and under each other are a mirror image of the vertical and horizontal flowers, respectively.

15 *010*
100 is always converted into 0.

16 *Lath 11*
All red laths have an odd number of holes and all blue laths have an even number.

17 *15*
The number equals the total number of sides of all geometric figures on the panel.

18 *2C*

19 *Honeybee 2*
This honeybee buzzes one fewer i in the first stanza.

20 *515253*
Surfers on the same team have a number that forms an ascending series across all the sails. For example, the white team has the numbers 79-80-81, 82-83-84, and 85-86-87.

Solutions

21 *Row 3*
This row also has four consecutive letters mixed around.

22 *Electron orbit B*
All electrons of the other orbits are sitting on the outer orbit.

23 *A=0, B=1*
The number of spots in each row alternate between a descending (3-2-1-0) and ascending (0-1-2-3-4-5-6) series of numbers.

24 *dolor*
All five-letter words are underlined.

25 *7*
The sum of the digits of the five numbers per rank always equals ten.
6+1+1+1+1=3+7+0+0=5+0+3+2+0=1+0+0+0+9=10

26 *3 2 7*
6 5 4
9 8 1

27 *I*
The vowels A, E, I, O, and U follow each other in the sentence.

28 *The lightning bolt in the middle.*

29 *214*
The next distance always equals the previous distance plus the final number in that distance.
187+7=194, 194+4=198, 198+8=206, 206+6=212, and 212+2=214.

30 *Piece C*
The seal only rests on pentagonal pieces of ice.

Solutions

31 *E5*
There is a golf ball equidistant from each pair of flags, both vertically and horizontally.

32 *Rabbit 5*
Only the rabbits that are awake show their teeth. Those who are asleep do not.

33 *8*
On each firework there is one more blue star than red, and one more red star than yellow.

34 *Group C*
The two red matches are at the bottom of the other groups.

35 *Baseball cap 2*
The logo on this cap is a mirror image.

36 *2123*
In each lollipop pair, one color remains the same while the other two colors swap places.

37 *13+39+48=100*

38 *38*
The number on a token always equals the sum of the last number in that row and that column. 6+32=38

39 *14*
20=14+0+1+2+3

40 *1569*
The following calculations are always performed with the two first numbers:
Subract the second number from the first (3-2=1)
Add the second number to the first (3+2)
Multiply the first number by the second (3x2=6)
Raise the first number to the power of the second number ($3^2=9$).

Solutions

41 *Three cheese cubes in the lower right corner.*

42 *Worm C*
The order of the body part colors in increasing dominance is blue-white-yellow-red.

43 *Flower branch 6*
On all the other branches, the flowers have as many petals as there are flowers on the branch.

44 *343 degrees*
28 degrees is always added to the degree of the angle.
The question mark is at 315 degrees. 315+28=343 degrees

45 *5*
The number after the decimal point is equal to the number of letters in the name on the sugar packet.

46 *Yogurt container 5*
The spiral on the cover is a mirror image.

47 *Point 7*

48 *Stamp 4*
The stamps touch each other according to the following color pairs:
yellow–purple, black–white, blue–orange, and red–green.

49 *5*
The sum of the digits of every number forms an ascending series from 5 to 9.

50 *Buoy 4*
The skipper changes direction according to the color of buoy he passes.
Red=90 degrees, yellow=270 degrees, and black=0 degrees.
This is the path that he follows:
12-11-10-7-6-9-8-5-3-1-2.

Solutions

51 *Cube 4F*
Starting from a red cube, odd numbers run horizontal and even number run vertical.

52 *Resistor A*
Resistors with a red band are placed on the circuit board horizontally; those with a blue band are vertical.

53 *OR*
The name of all the red documents consists of consecutive letters that are three places farther in the alphabet.
For the white documents, the letters are two places farther.

54 *5 blocks*
Blocks of the same color always form a square, so you need three more white and two more red blocks.

55 *Square 4*

56 *Group C has a symbol with five lines.*
All the other groups consist of symbols with one, two, three, and four lines.

57 *Symbol 2*
If you stack all the letters per word on top of each other, the non-overlapping black blocks form the symbol. For symbol 2, the square in the second column and second row should be empty.

58 *7*
The digit in the bottom row is always one higher than the digit in the top row.

59 *3*
In the zones with coins, there is always one coin less than the number of angles in the zone.

60 *Star 8*
All the other stars are reflected in the water.

Solutions

61 Shape 2
Green is dominant in E and F.
In the previous series it emerged that the square was dominant over the diamond, but the triangle was dominant over the square. The order in ascending superiority is: D–E–B–A–C.

62 Figure 5 is shown three times as a three-in-a-row.

63
1 HURDLES–JAVELIN
2 PARTNER–WEDDING
3 BOLERO–FOXTROT

64

1	*ATLANTIC*	*–N*	*CATTAIL*
2	*CAMPAIGN*	*–A*	*CAMPING*
3	*HAIRLINE*	*–I*	*INHALER*
4	*BERMUDAS*	*+T*	*DRUMBEATS*
5	*BACHELOR*	*–B*	*CHOLERA*
6	*EARPHONE*	*+C*	*CHAPERONE*
7	*INCREASE*	*–A*	*SINCERE*

65
1 LUXURY HOTEL
2 LEMON SORBET
3 SWIMSUIT
4 CAMERA
5 RIVER CRUISE

66
1 G-string
2 Surrounded
SUR round ED
3 Middleweight
WEIGHT WEIGHT WEIGHT WEIGHT WEIGHT
4 Bind
b in D

67 IS
All the other letters form pairs of two consecutive letters in the alphabet.

Solutions

68	*Several solutions are possible:*			
	1	*chocolate–gift–birthday–date–calendar–time–stress*		
	2	*shepherd–sheep–stew–kitchen–chef–hat–winter*		
	3	*luck–lottery–money–bank–robbery–criminal–prison*		
	4	*car–accident–hospital–surgery–anesthesia–dreaming–awake*		
	5	*naked–clothes–warm–sweating–sauna–steam–train*		
	6	*football–goal–score–board–lesson–teacher–school*		
69	*1*	*Airline*		
	2	*Countdown*		
	3	*Sunspots*		
		Sun + spots		
	4	*Artwork*		
		AR two RK		
70	*1*	*NEBRASKA*	*–A*	*BANKERS*
	2	*CANOEING*	*+R*	*IGNORANCE*
	3	*GRENADES*	*+E*	*RENEGADES*
	4	*BASELINE*	*–E*	*LESBIAN*
	5	*COMPLETE*	*–P*	*TELECOM*
	6	*FACELIFT*	*–E*	*AFFLICT*
	7	*CHARISMA*	*+R*	*ARMCHAIRS*
71	*1*	*PRISON*		
	2	*SEX CHANGE*		
	3	*CHEATING*		
	4	*IMBECILE*		
	5	*OVERWEIGHT*		
72		*A*		
	1	*AD*		
	2	*LAD*		
	3	*LAID*		
	4	*IDEAL*		
	5	*LADIES*		
	6	*MISLEAD*		
	7	*IDEALISM*		

Solutions

73	*1*	*PROCESS–SUSPECT*		
	2	*CHOPPER–SCOOTER*		
	3	*DESPAIR–PASSION*		
74	*1*	*APES–GIBBER*		
	2	*ELEPHANTS–TRUMPET*		
	3	*COCKS–CROW*		
	4	*DOGS–BARK*		
	5	*FLIES–BUZZ*		
	6	*HAWKS–SCREAM*		
	7	*PENGUINS–HONK*		
75	*1*	*BULLS–BELLOW*		
	2	*MOSQUITOES–WHINE*		
	3	*SNAKES–HISS*		
	4	*TURKEYS–GOBBLE*		
	5	*PARROTS–TALK*		
	6	*FROGS–CROAK*		
	7	*SWANS–CRY*		
76	*1*	*FLYING*		
	2	*SPEED*		
	3	*WOMEN*		
	4	*NEW THINGS*		
	5	*WORKING*		
77	*1*	*LOCATION*	*+I*	*COALITION*
	2	*BEDCOVER*	*–B*	*COVERED*
	3	*DOORSTEP*	*–S*	*TORPEDO*
	4	*COLLAPSE*	*–O*	*SCALPEL*
	5	*KINGFISH*	*–K*	*FISHING*
	6	*DIAMETER*	*+G*	*EMIGRATED*
	7	*NAPOLEON*	*+M*	*MONOPLANE*

78 *Group 2*
In the other groups, the same letters are always across from each other.

Solutions

79	*1*	*POLITICIAN*		
	2	*PSYCHIATRIST*		
	3	*HUNCHBACK*		
	4	*MARRIAGE*		
	5	*BUNGEE JUMP*		
80		*F*		
	1	*FA*		
	2	*FAT*		
	3	*FAST*		
	4	*FACTS*		
	5	*CRAFTS*		
	6	*FACTORS*		
	7	*FORECAST*		
81	*1*	*OPERATOR*	*–A*	*TROOPER*
	2	*TRIDENTS*	*–R*	*DENTIST*
	3	*DISAGREE*	*+T*	*TRAGEDIES*
	4	*NICOTINE*	*+F*	*INFECTION*
	5	*FIREBALL*	*–F*	*BRAILLE*
	6	*INCUBATE*	*–U*	*CABINET*
	7	*MAGAZINE*	*–E*	*AMAZING*
82	*1*	*Reduce*		
		red UCE		
	2	*Fats Domino*		
		fat S domino		
	3	*Monkey*		
		M on key		
	4	*Seventies*		
		seven ties		
83	*1*	*ASSAULT–PURSUIT*		
	2	*COTTON–LEATHER*		
	3	*BIOLOGY–TEACHER*		

Solutions

84 1 *GEOGRAPHY*
2 *ENGLISH*
3 *CHEMISTRY*
4 *HISTORY*
5 *MATHEMATICS*
6 *ECONOMICS*

85 *No solution*

86 *No solution*

87 *Set 2*

88 *No solution*

89 1 *Photos B and D*
2 *Photo C*
3 *Photo F*
4 *Photo A (The NYC was deleted from the door of the cab.)*

90 1 *The number 7 was added in the text*
2 *The square was rotated*
3 *The hexagon was rotated*
4 *The number 7 is missing an upper left piece*
5 *The opening is missing on the letter R*

91 *No solution*

92 1 *menhir sculptor*
2 *in Peru*
3 *North Pole*
4 *Rubicon*
5 *oenology*
6 *acronym*
7 *the 44th*
8 *braille*
9 *Spiderman*
10 *Polaris*

Solutions

93 *No solution*

94 *No solution*

95 *No solution*

96 *1 mackerel*
2 herring
3 sardine
4 salmon
5 tuna

97 *The die in the third row and the third column is wrong. On the right, left, and top side there are one, two, and three spots respectively.*

98 *Tile 7*

99 *1 45*
2 No solution
Note: The sum of identically colored numbers is always ten.

100 *1 41,976*
2 5,555

101 *Cube 2*
The three sides of the large cube are colored identically.

102 *Rider 5 will reach the finish line in seven seconds.*

103 *Back B*

104 *Angle 3*
On this view from above, the parts on the back should be reversed.

105 *Piece C*

Solutions

106 *Shapes 4, 5, and 6*

107 *Stamp B*
The square in the top left-hand corner has been rotated forty-five degrees.

108 *Surface 2*
Surface 2 should be pink. All upper surfaces are yellow, all side surfaces are pink, and all frontal surfaces are red.

109 *Gift 3*
To be identical to the other gifts, the white triangle must be under the black triangle.

110 *Profile D*
The cube with the white top face must be in front of the cube with the purple top face.

111 *Piece 4*
Only pieces 2, 4, and 6 consist of the ten blocks that you need to complete the cube. Only piece 4 has the right shape and the right colors.

112 *Angle 4*
The chain of the drawbridge is too long.

113 *Group C*
In all the other pairs, the two signs form a completely colored circle without overlapping when stacked on top of each other.

114 *Strip 3*
All the other strips form a right angle when unfolded.

115 *Cube 2*
The colors of the tops alternate between red and yellow per level, and the colors on the sides alternate between black and white. Only cube 2 contains yellow, black, and white.

Solutions

116 *G*
The GPS always goes straight.
The route goes through the following roundabouts:
5–3–4–1–3–5–7–2–1–4–6.

117 *74735*
First number=numbers rotate from front,
top, and side, advancing one block each time.
Second number=numbers rotate from top,
side, and front, advancing one block each time.
Third number=numbers rotate from side,
front, and top, advancing one block each time.

118 *Block 3*

119 *Shape 4*

120 *A2*
Stack A belongs in location 2.
Per stack, the colors and the shapes shift down two places.

121 *Piece 3*
To be useful, the top purple square of piece 3 must shift left.

122 *Pieces 1, 5, and 6 make a rectangle.*

Solutions

123

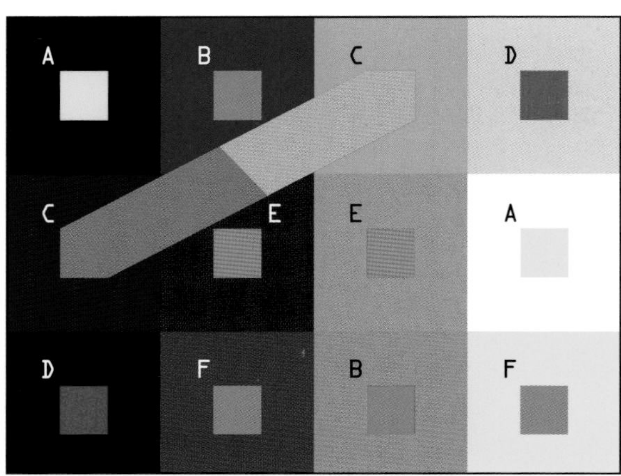

124

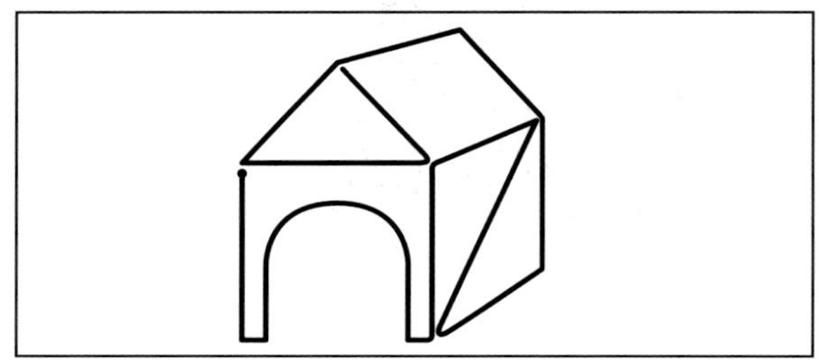

125

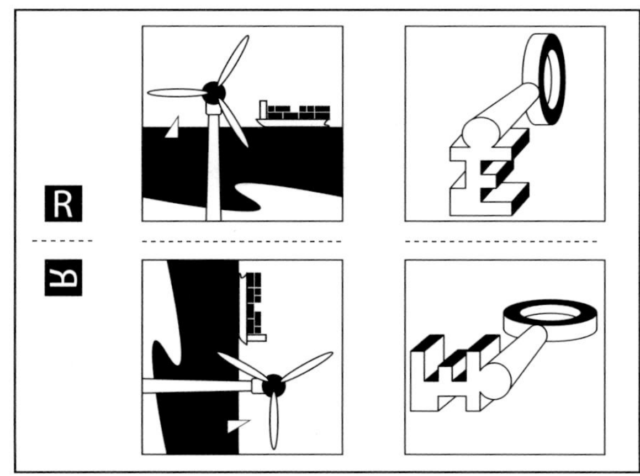

Solutions

126

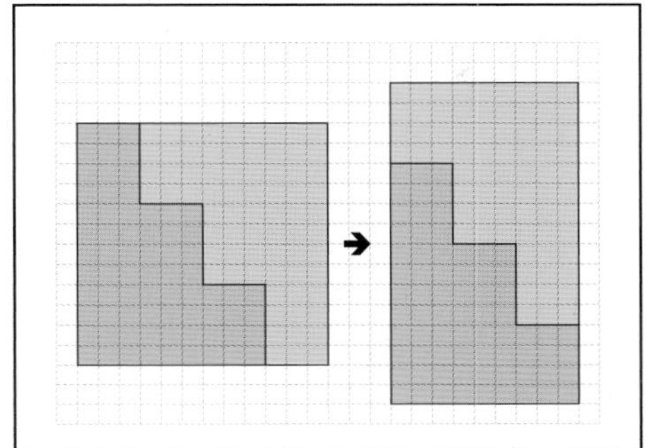

127

128

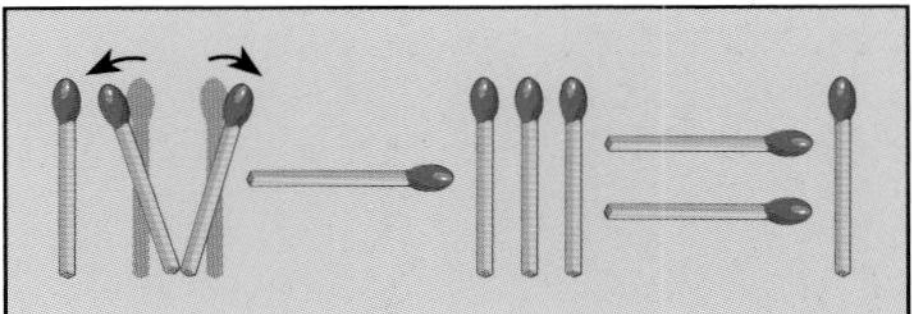

129 *No solution*

130 *V*
The phrase "Seeing is believing" appears when you read all the numbers as letters.

131 *Groups 2 and 6*
All other groups only have squares.

Solutions

132

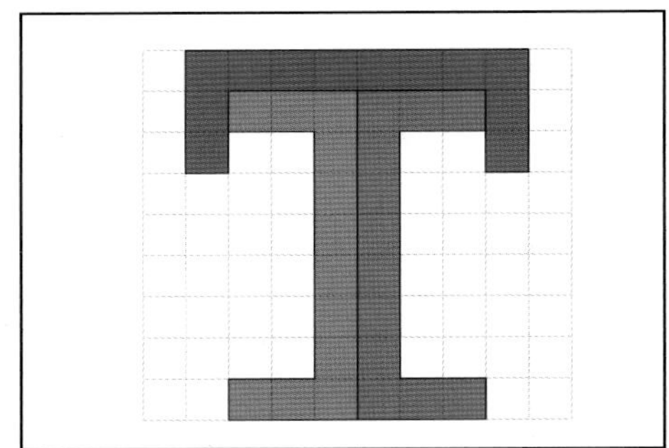

133	*1*	*confident–proud–modest–stubborn–amused*
	2	*perplexed–shocked–astonished–surprised–panicked*
	3	*doubt–worried–dejected–discouraged–humiliated*
134	*4*	*contemptuous–tetchy–angry–impudent–aggressive*
	5	*reminiscent–jealous–skeptical–conceited–reserved*

135

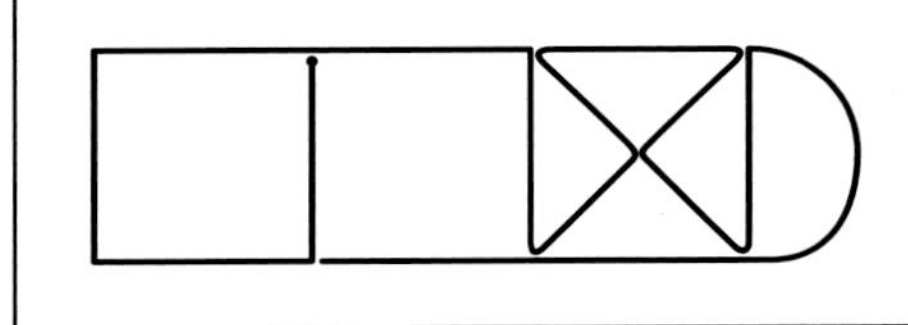

136

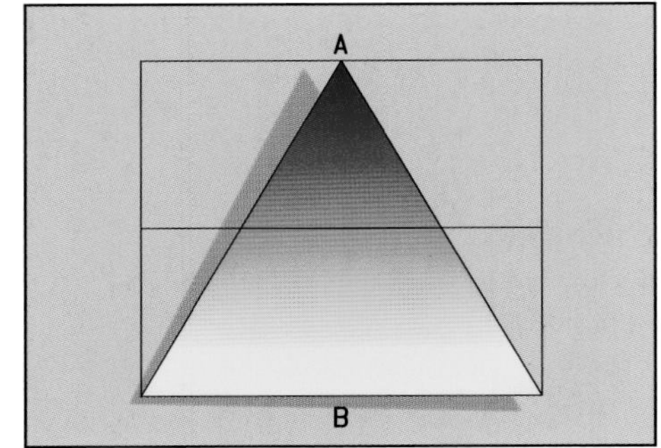

137 *23*

Solutions

138

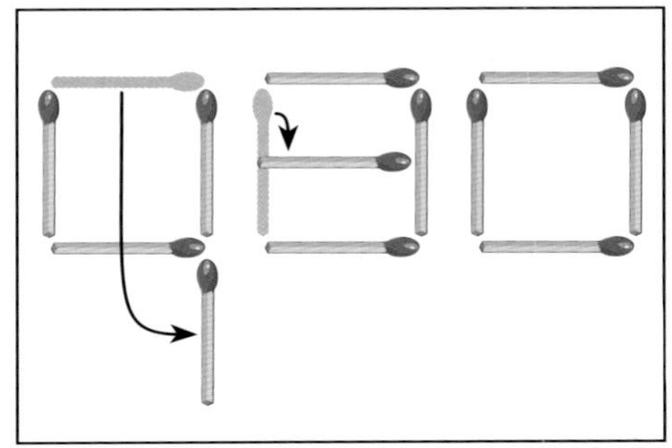

139 *Today*

140 *No solution*

141

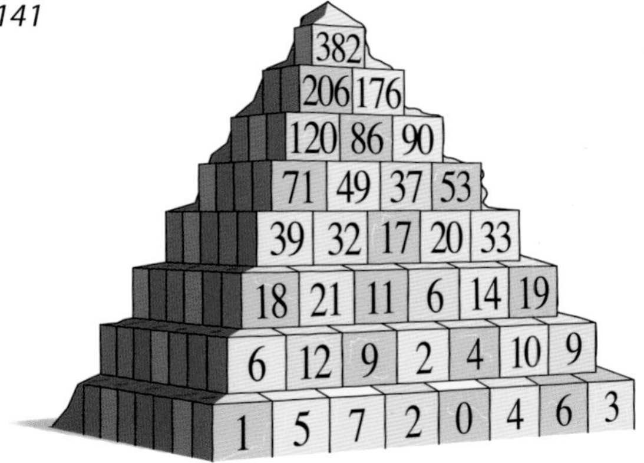

You can only make 6 with 1+5,
and not with 2+4 because you can only make 9 with 2+7,
and not with 6+3 because 6 is already shown,
and not with 4+5 otherwise you cannot make 6.
The 2 is on the right under the 9 because the 7 is a higher value than the 6 on the second level.

Solutions

142

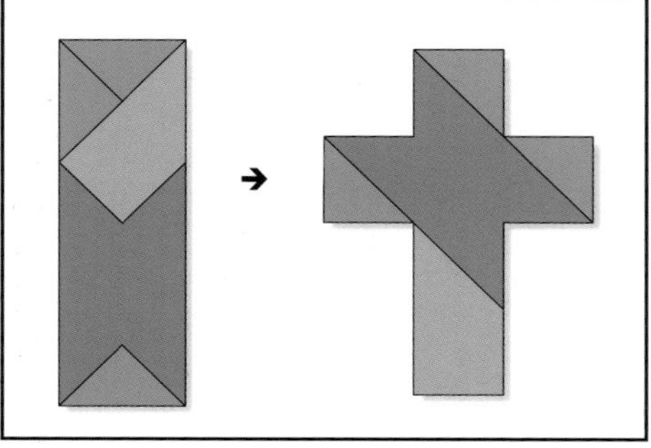

143 *21*
The next number always equals the sum of the two previous numbers in the Fibonacci series.

144

145 *No solution*

146

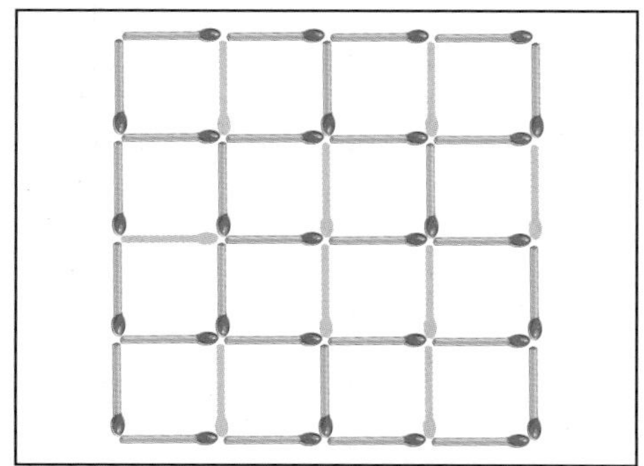

Solutions

Number Clusters

147

5	5	5	5	**5**	7
8	8	8	**8**	**1**	7
8	**6**	6	8	**7**	7
2	**2**	6	8	**3**	7
4	**4**	6	**8**	3	**7**
4	4	6	**6**	3	7

148

6	6	6	6	6	**6**
5	5	5	5	**5**	8
4	4	4	**4**	8	8
3	3	**3**	7	7	8
2	**2**	**7**	7	**7**	8
1	7	7	**8**	8	8

149

8	8	2	**1**	7	6
4	8	**2**	5	**7**	6
4	8	5	**5**	7	6
4	8	5	5	7	6
4	8	3	**3**	7	6
8	8	3	**7**	7	**6**

150

4	4	8	**5**	5	5
4	**8**	8	8	8	5
4	7	**7**	7	8	**5**
6	6	6	7	**8**	8
3	**3**	6	7	**2**	2
3	6	6	7	7	**1**

151

8	8	8	8	8	2
1	8	8	8	3	**2**
7	7	7	**7**	3	**3**
7	7	4	4	4	**4**
7	5	5	5	5	**5**
6	6	6	6	6	**6**

Solutions

152

5	5	5	3	3	3
5	5	8	8	8	2
6	6	6	7	8	2
6	7	7	7	8	1
6	7	8	8	8	4
6	7	7	4	4	4

153

6	6	6	6	6	6
8	5	5	5	5	5
8	8	4	4	4	4
8	7	7	3	3	3
8	7	7	7	2	2
8	8	8	7	7	1

154

7	7	7	6	6	5
7	7	7	6	6	5
7	8	8	6	6	5
2	2	8	1	3	5
4	4	8	3	3	5
4	4	8	8	8	8

155

6	6	6	6	6	6
8	8	8	8	8	8
4	4	4	3	3	8
4	2	2	1	3	8
7	7	7	7	7	7
5	5	5	5	5	7

156

7	4	4	2	2	6
7	7	4	4	6	6
3	7	7	7	6	1
3	3	7	6	6	8
5	5	5	5	5	8
8	8	8	8	8	8

157

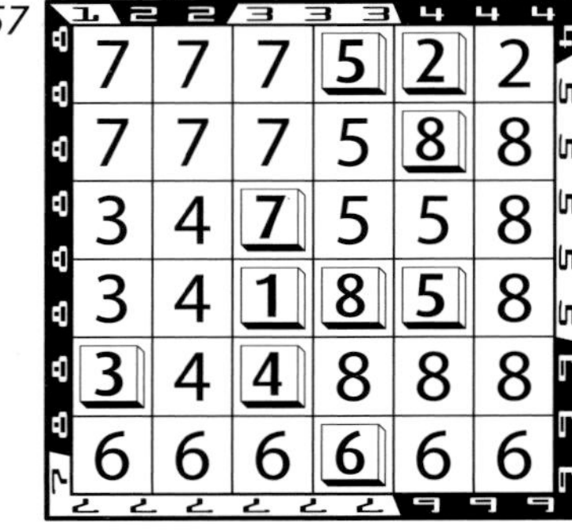

7	7	7	5	2	2
7	7	7	5	8	8
3	4	7	5	5	8
3	4	1	8	5	8
3	4	4	8	8	8
6	6	6	6	6	6

Solutions

Sunny Weather Forecast

158

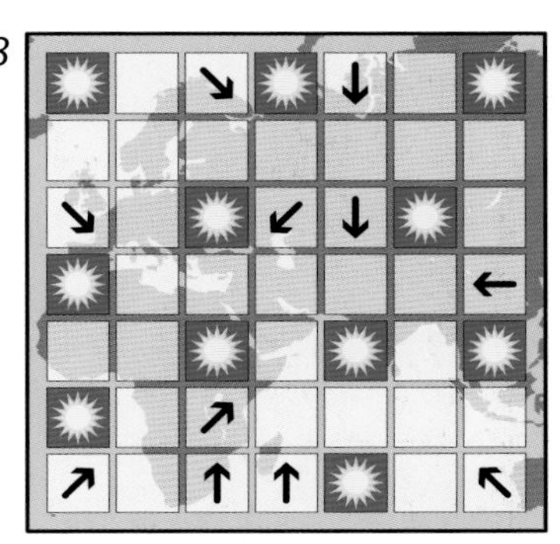

159

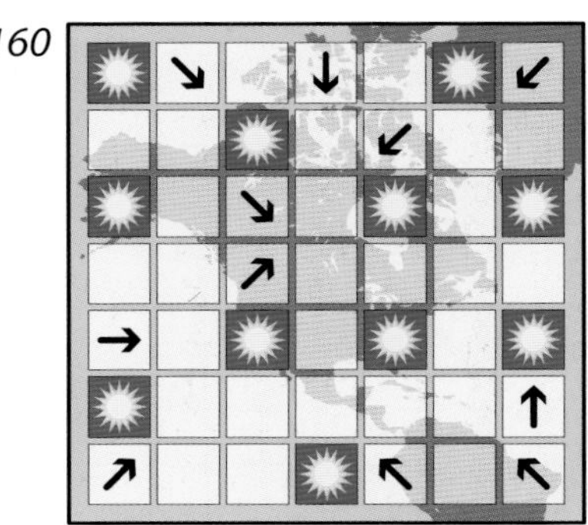

160

161

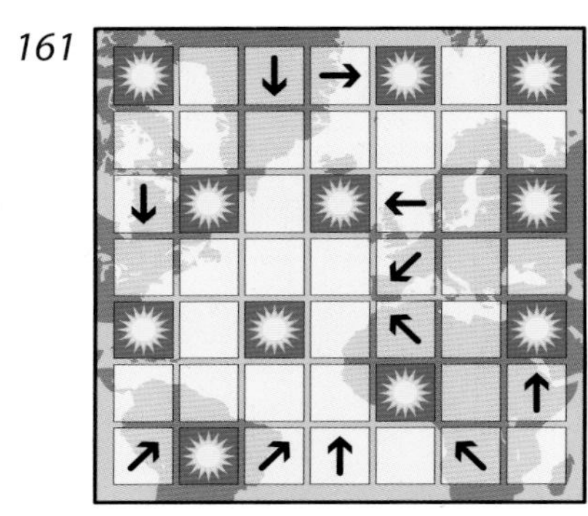

162

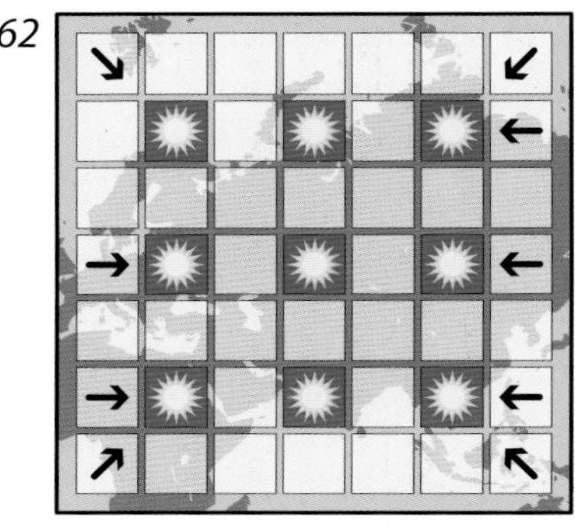

Solutions

163

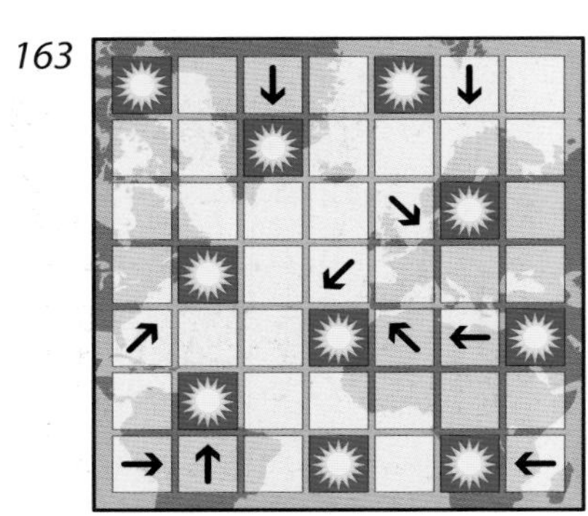

164

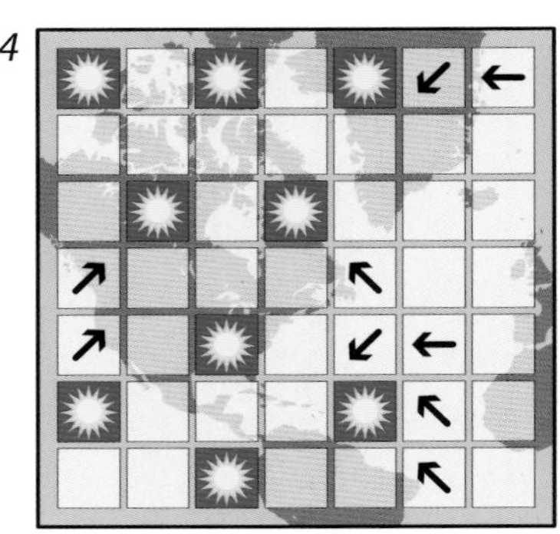

165

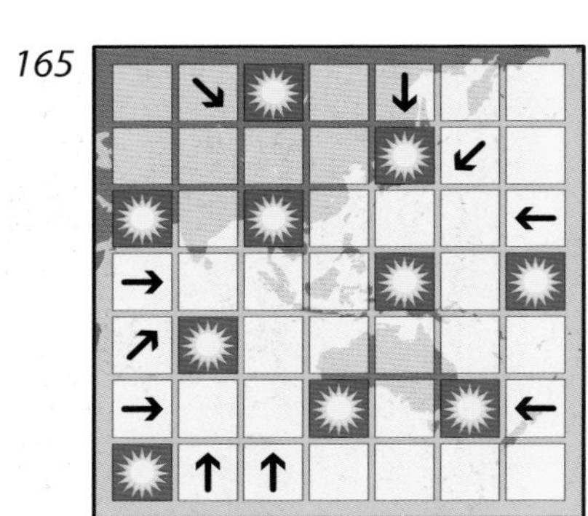

166

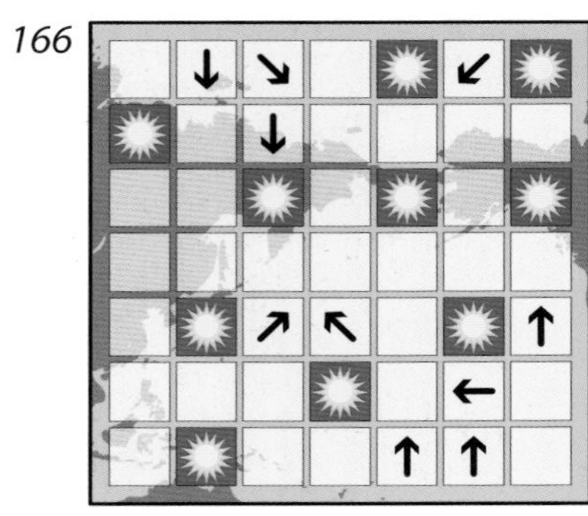

167

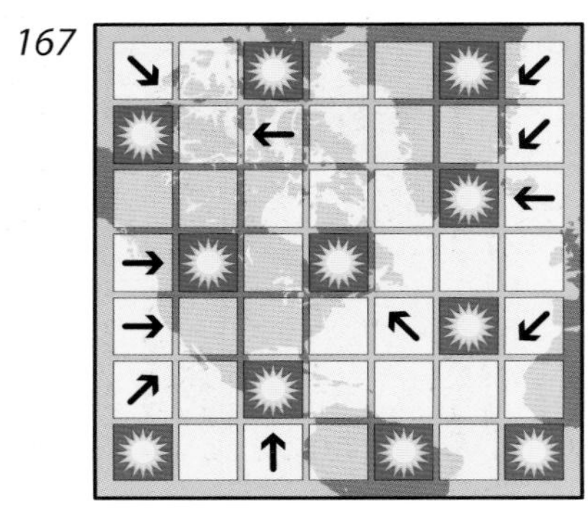

168

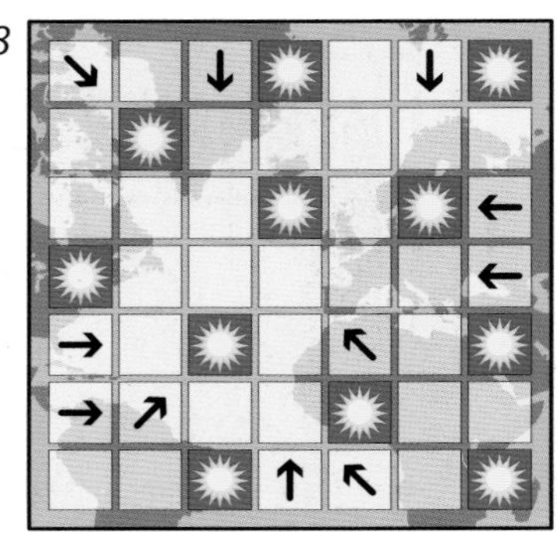

Solutions

Continuous Line

169

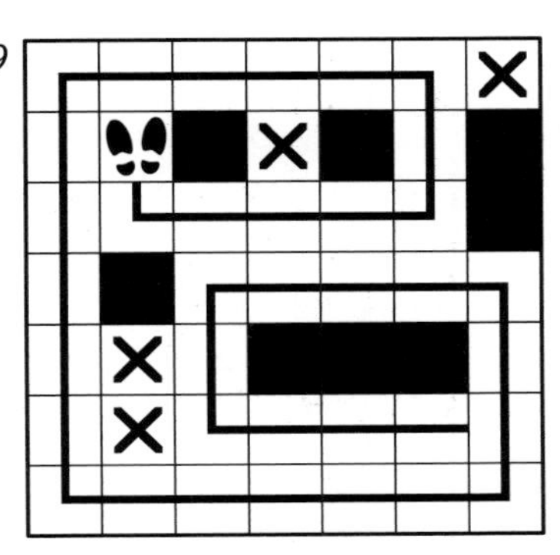

170

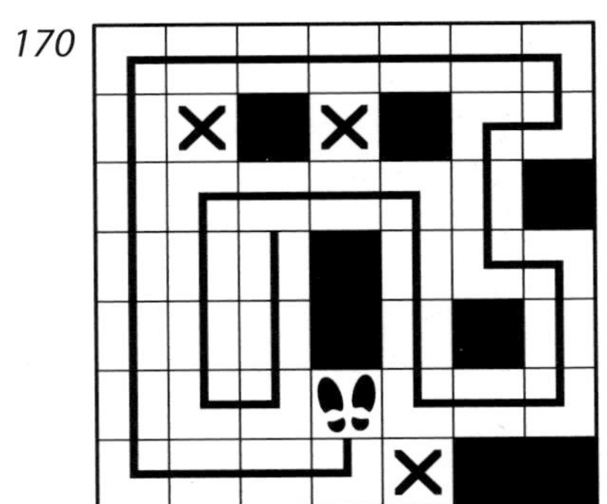

171

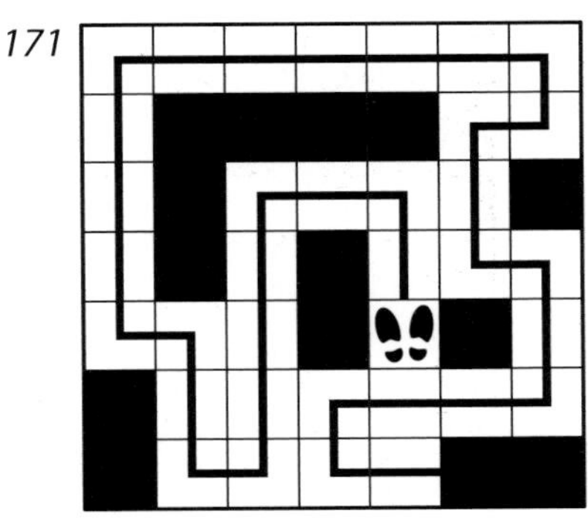

172

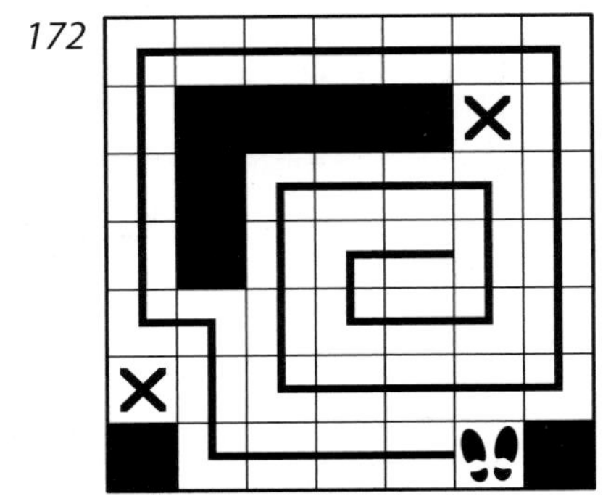

173

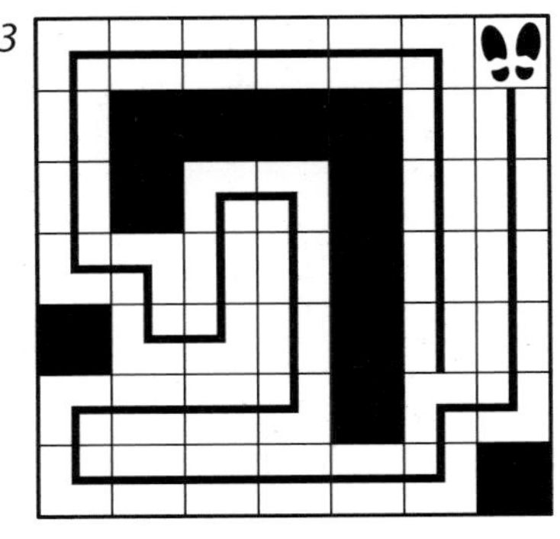

Solutions

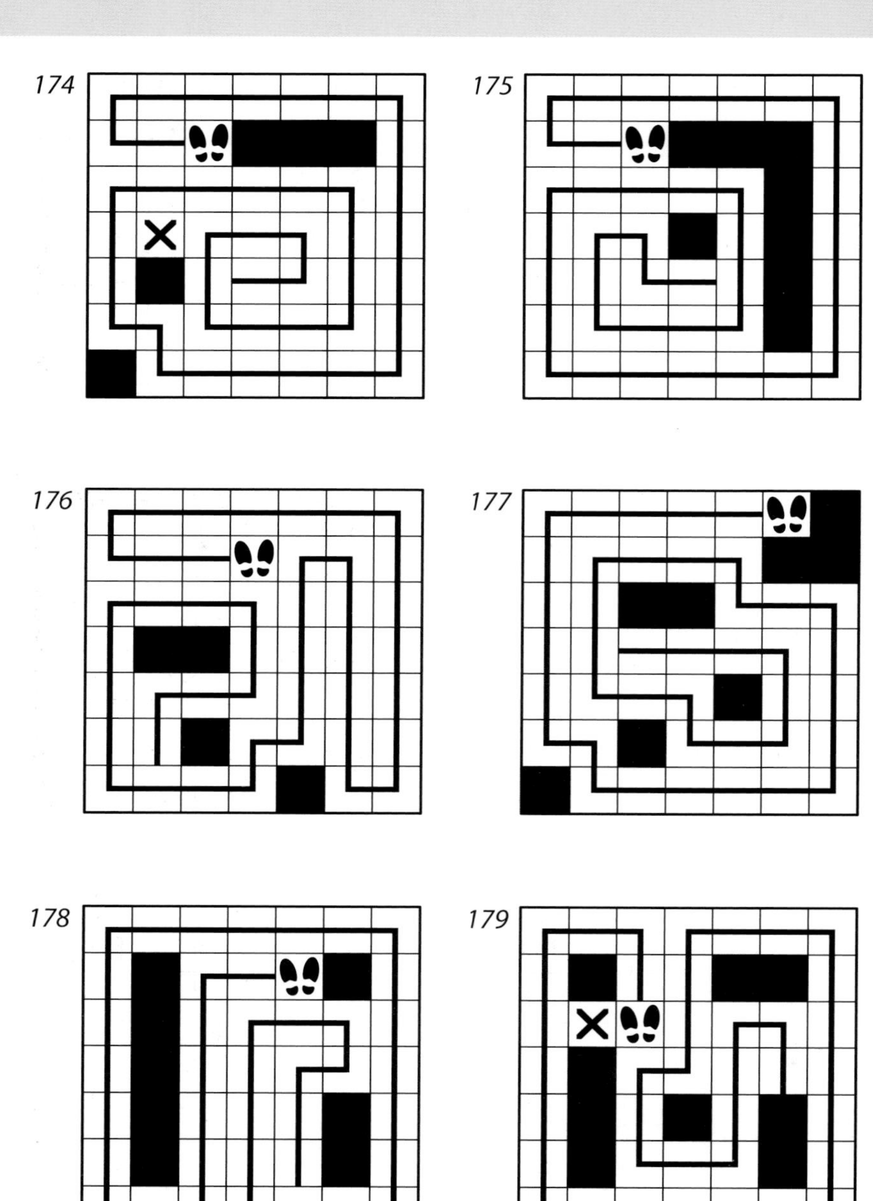

Solutions

Football & Golf

180

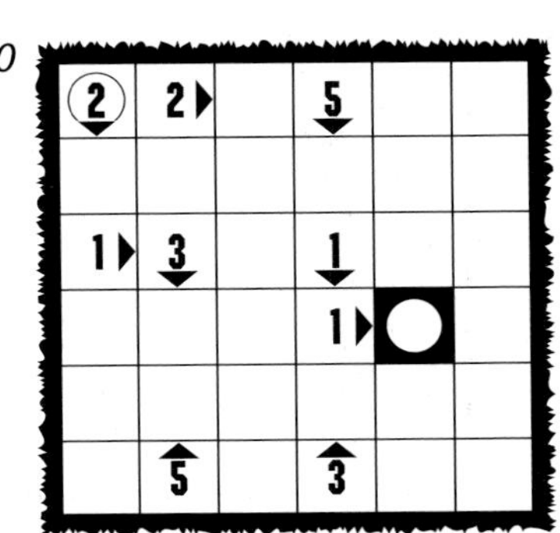

181

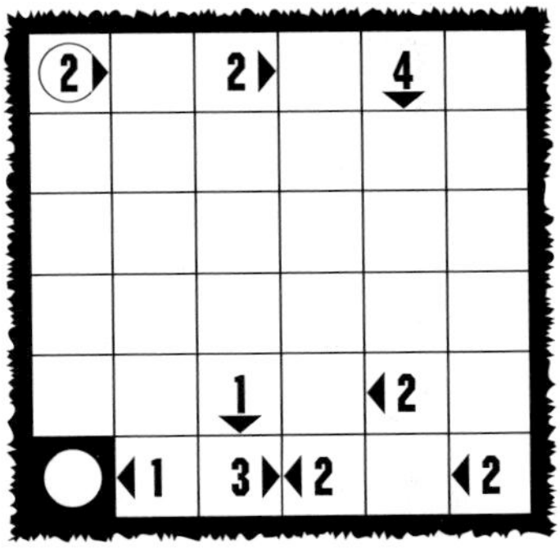

182

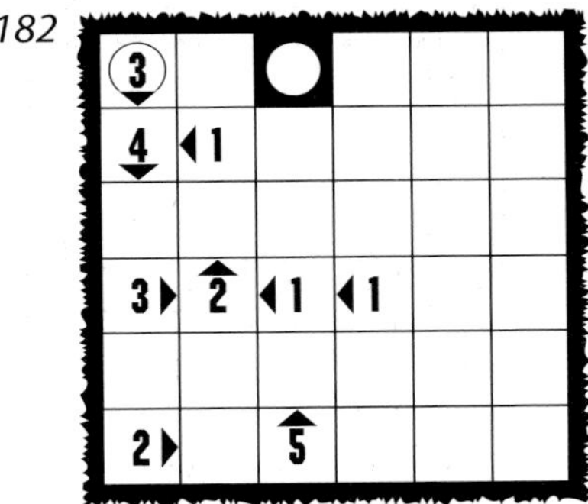

183

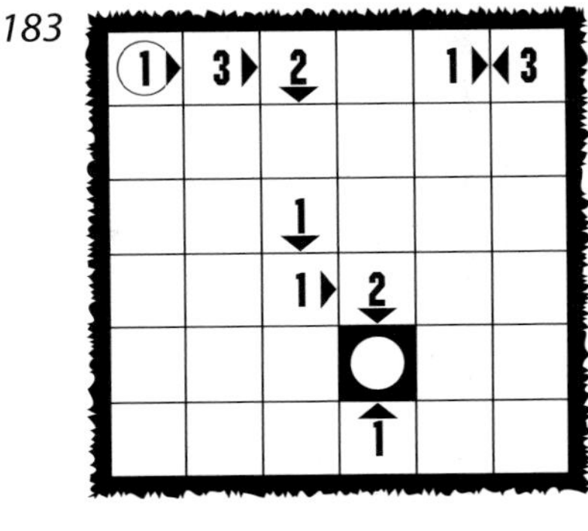

184

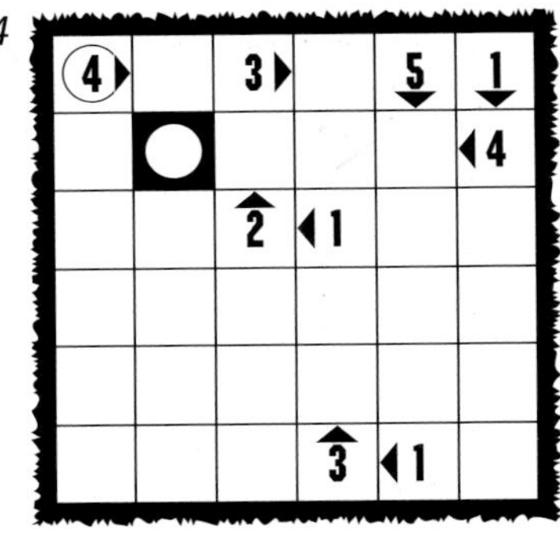

Solutions

185

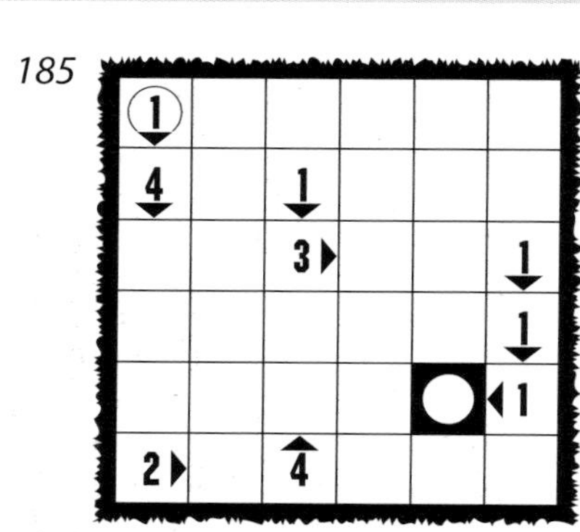

186

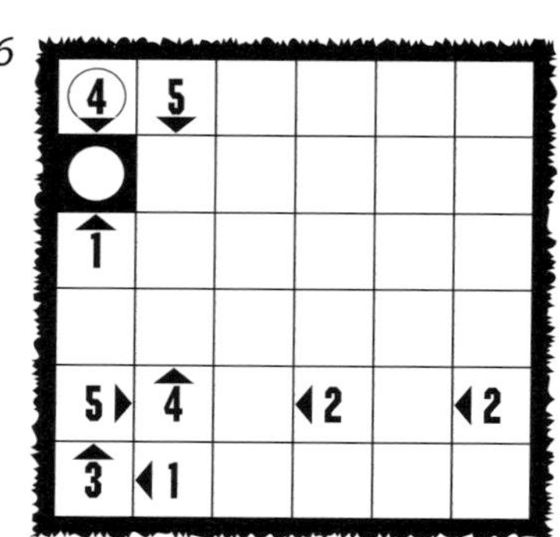

187

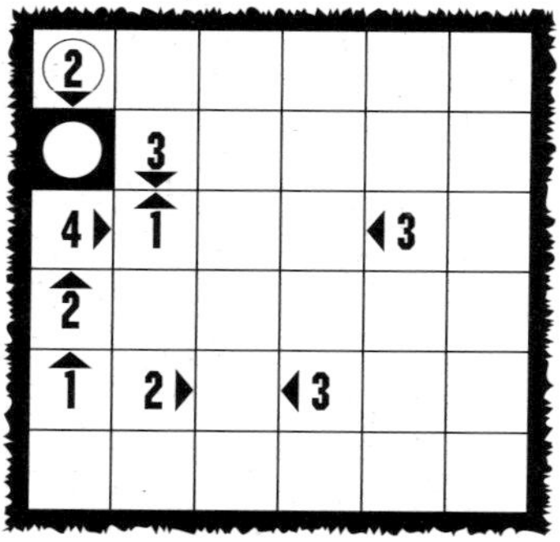

(2)↓					
●	3↓				
4▶	↑1			◀3	
↑2					
↑1	2▶		◀3		

188

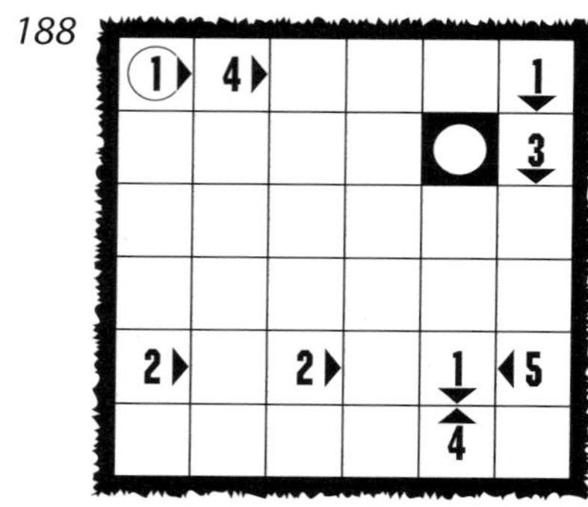

189

(4)▶	4↓			◀3	
					●
	4▶	◀1			↑1
	4▶	↑1			◀3

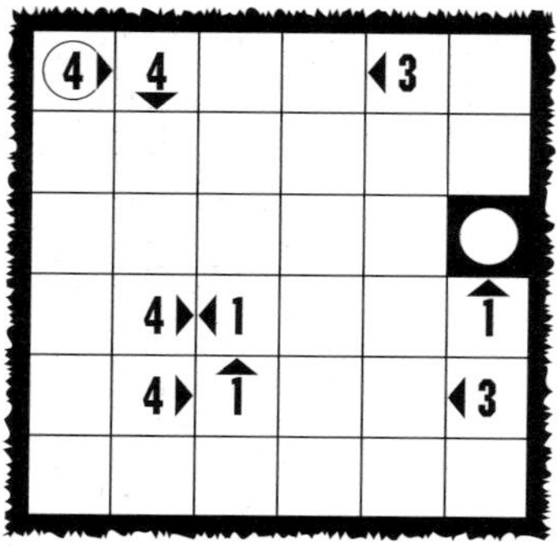

190

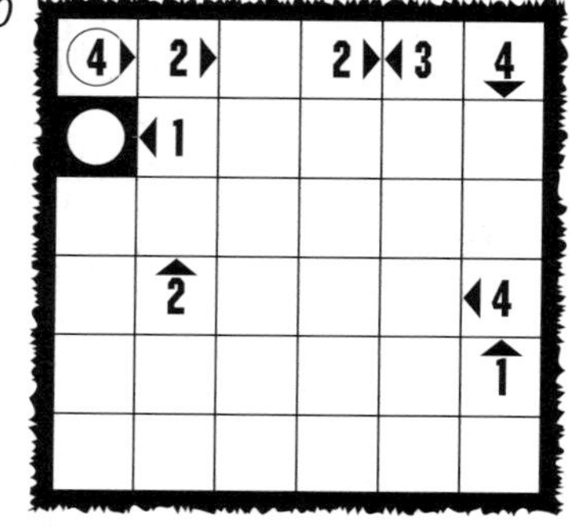

Solutions

Binary

191

0	1	1	0	1	0	0	1	0	0	1	1
1	1	0	0	1	0	1	0	1	1	0	0
1	0	1	1	0	1	0	0	1	1	0	0
0	1	0	1	0	1	0	1	0	0	1	1
0	0	1	0	1	0	1	1	0	1	0	1
1	1	0	1	0	0	1	0	1	0	1	0
0	0	1	1	0	1	0	0	1	0	1	1
1	0	1	0	1	0	1	1	0	1	0	0
1	1	0	0	1	1	0	0	1	0	0	1
0	0	1	1	0	0	1	1	0	1	1	0
1	0	0	1	0	1	1	0	1	1	0	0
0	1	0	0	1	1	0	1	0	0	1	1

False: Wrestling was originally a Greek-Roman sport.

192

0	1	0	1	0	0	1	1	0	1	1	0
0	1	0	1	0	1	1	0	1	1	0	0
1	0	1	0	1	0	0	1	1	0	0	1
1	1	0	1	0	0	1	0	0	1	1	0
0	0	1	0	1	1	0	1	0	1	1	0
1	0	1	1	0	0	1	0	1	0	0	1
0	1	0	0	1	1	0	1	0	1	0	1
1	0	0	1	1	0	0	1	1	0	1	0
0	1	1	0	0	1	1	0	1	0	0	1
1	0	1	0	1	0	1	0	0	1	1	0
1	0	0	1	0	1	0	1	0	0	1	1
0	1	1	0	1	1	0	0	1	0	0	1

False: Bakelite is an early plastic, invented by the Belgian chemist Leo Baekeland.

193

1	0	0	1	0	0	1	0	1	1	0	1
0	1	0	1	0	1	0	0	1	1	0	1
1	0	1	0	1	1	0	1	0	0	1	0
0	1	1	0	1	0	1	0	1	0	1	0
1	0	0	1	0	1	1	0	0	1	0	1
0	1	0	1	1	0	0	1	1	0	1	0
1	0	1	0	1	1	0	1	0	1	0	0
1	1	0	1	0	0	1	0	1	0	0	1
0	0	1	0	1	1	0	1	0	1	1	0
1	1	0	0	1	0	1	1	0	1	0	0
0	0	1	1	0	0	1	0	1	0	1	1
0	1	1	0	0	1	0	1	0	0	1	1

False: The Nile is the longest river.

194

0	1	0	0	1	1	0	1	1	0	1	0
1	0	1	0	1	0	1	0	0	1	1	0
1	1	0	1	0	0	1	0	1	0	0	1
0	0	1	1	0	1	0	1	0	1	1	0
1	1	0	0	1	0	1	1	0	0	1	0
0	0	1	0	1	1	0	0	1	1	0	1
0	0	1	1	0	0	1	0	1	1	0	1
1	1	0	1	0	1	0	1	0	0	1	0
1	1	0	0	1	1	0	0	1	0	0	1
0	0	1	1	0	0	1	1	0	1	0	1
1	1	0	1	0	0	1	0	0	1	1	0
0	0	1	0	1	1	0	1	1	0	0	1

True

195

0	1	1	0	1	0	0	1	1	0	1	0
1	1	0	1	0	1	1	0	0	1	0	0
0	0	1	1	0	1	0	1	1	0	0	1
1	1	0	0	1	0	1	1	0	0	1	0
0	0	1	1	0	1	1	0	0	1	0	1
0	0	1	1	0	1	0	0	1	1	0	1
1	1	0	0	1	0	0	1	1	0	1	0
1	0	0	1	1	0	1	0	0	1	1	0
0	1	1	0	0	1	0	1	1	0	0	1
1	0	0	1	1	0	1	0	0	1	0	1
1	1	0	0	1	0	1	0	0	1	1	0
0	0	1	0	0	1	0	1	1	0	1	1

False: It was invented by Bíró in 1938.

Solutions

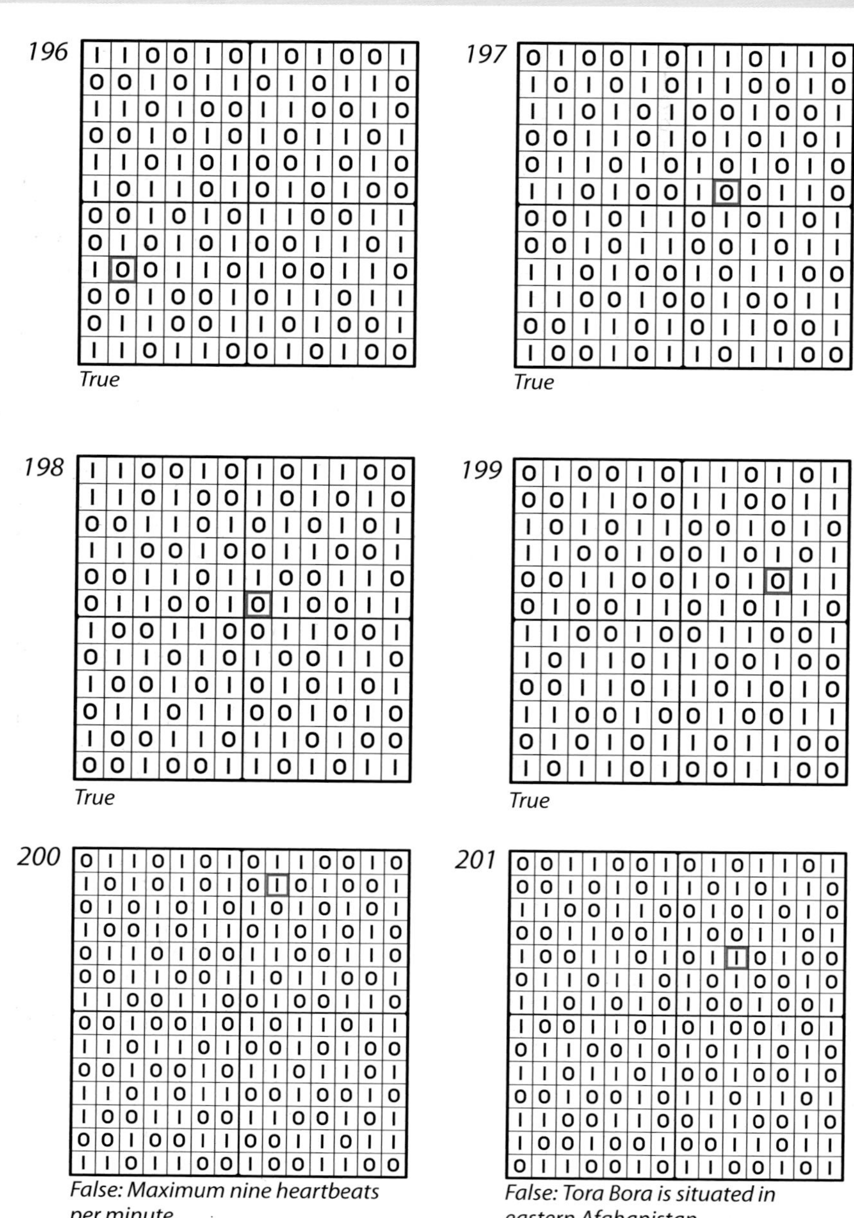

196

I	I	O	O	I	O	I	O	I	O	O	I
O	O	I	O	I	I	O	I	O	I	I	O
I	I	O	I	O	O	I	I	O	O	I	O
O	O	I	O	I	O	I	O	I	I	O	I
I	I	O	I	O	I	O	O	I	O	I	O
I	O	I	I	O	I	O	I	O	I	O	O
O	O	I	O	I	O	I	I	O	O	I	I
O	I	O	I	O	I	O	O	I	I	O	I
I	O	O	I	I	O	I	O	O	I	I	O
O	O	I	O	O	I	O	I	I	O	I	I
O	I	I	O	O	I	I	O	I	O	O	I
I	I	O	I	I	O	O	I	O	I	O	O

True

197

O	I	O	O	I	O	I	I	O	I	I	O
I	O	I	O	I	O	I	I	O	O	I	O
I	I	O	I	O	I	O	O	I	O	O	I
O	O	I	I	O	I	O	I	O	I	O	I
O	I	I	O	I	O	I	O	I	O	I	O
I	I	O	I	O	O	I	O	O	I	I	O
O	O	I	O	I	I	O	I	O	I	O	I
O	O	I	O	I	I	O	O	I	O	I	I
I	I	O	I	O	O	I	O	I	I	O	O
I	I	O	O	I	O	O	I	O	O	I	I
O	O	I	I	O	I	O	I	I	O	O	I
I	O	O	I	O	I	I	O	I	I	O	O

True

198

I	I	O	O	I	O	I	O	I	I	O	O
I	I	O	I	O	O	I	O	I	O	I	O
O	O	I	I	O	I	O	I	O	I	O	I
I	I	O	O	I	O	O	I	I	O	O	I
O	O	I	I	O	I	I	O	O	I	I	O
O	I	I	O	O	I	O	I	O	O	I	I
I	O	O	I	I	O	O	I	I	O	O	I
O	I	I	O	I	O	I	O	O	I	I	O
I	O	O	I	O	I	O	I	O	I	O	I
O	I	I	O	I	I	O	O	I	O	I	O
I	O	O	I	I	O	I	I	O	I	O	O
O	O	I	O	O	I	I	O	I	O	I	I

True

199

O	I	O	O	I	O	I	I	O	I	O	I
O	O	I	I	O	O	I	I	O	O	I	I
I	O	I	O	I	I	O	O	I	O	I	O
I	I	O	O	I	O	O	I	O	I	O	I
O	O	I	I	O	O	I	O	I	O	I	I
O	I	O	O	I	I	O	I	O	I	I	O
I	I	O	O	I	O	O	I	I	O	O	I
I	O	I	I	O	I	I	O	O	I	O	O
O	O	I	I	O	I	I	O	I	O	I	O
I	I	O	O	I	O	O	I	O	O	I	I
O	I	O	I	O	I	I	O	I	I	O	O
I	O	I	I	O	I	O	O	I	I	O	O

True

200

O	I	I	O	I	O	I	O	I	I	O	O	I	O
I	O	I	O	I	O	I	O	I	O	I	O	O	I
O	I	O	I	O	I	O	I	O	I	O	I	O	I
I	O	O	I	O	I	I	O	I	O	I	O	I	O
O	I	I	O	I	O	O	I	I	O	O	I	I	O
O	O	I	I	O	O	I	I	O	I	I	O	O	I
I	I	O	O	I	I	O	O	I	O	O	I	I	O
O	O	I	O	O	I	O	I	O	I	I	O	I	I
I	I	O	I	I	O	I	O	O	I	O	I	O	O
O	O	I	O	O	I	O	I	I	O	I	I	O	I
I	I	O	I	O	I	I	O	O	I	O	O	I	O
I	O	O	I	I	O	O	I	I	O	O	I	O	I
O	O	I	O	O	I	I	O	O	I	I	O	I	I
I	I	O	I	I	O	O	I	O	O	I	I	O	O

False: Maximum nine heartbeats per minute.

201

O	O	I	I	O	O	I	O	I	O	I	I	O	I
O	O	I	O	I	O	I	I	O	I	O	I	I	O
I	I	O	O	I	I	O	O	I	O	I	O	I	O
O	O	I	I	O	O	I	I	O	O	I	I	O	I
I	O	O	I	I	O	I	O	I	I	O	I	O	O
O	I	I	O	I	I	O	I	O	I	O	O	I	O
I	I	O	I	O	I	O	I	O	O	I	O	O	I
I	O	O	I	I	O	I	O	I	O	O	I	O	I
O	I	I	O	O	I	O	I	O	I	I	O	I	O
I	I	O	I	I	O	I	O	O	I	O	O	I	O
O	O	I	O	O	I	O	I	I	O	I	I	O	I
I	I	O	O	I	I	O	O	I	I	O	O	I	O
I	O	O	I	O	O	I	O	O	I	I	O	I	I
O	I	I	O	O	I	O	I	I	O	O	I	O	I

False: Tora Bora is situated in eastern Afghanistan.

Solutions

202

1	0	1	0	1	1	0	1	1	0	0	1	0	0
0	1	0	1	0	1	0	1	0	1	1	0	0	1
0	0	1	1	0	0	1	0	0	1	1	0	1	1
1	1	0	0	1	0	0	1	1	0	0	1	1	0
0	0	1	1	0	1	0	1	0	0	1	1	0	1
1	1	0	0	1	0	**1**	0	1	1	0	0	1	0
0	0	1	0	0	1	0	1	1	0	1	0	1	1
1	1	0	1	0	0	1	0	0	1	0	1	0	1
1	1	0	1	1	0	1	0	1	0	0	1	0	0
0	0	1	0	1	1	0	1	0	1	1	0	1	0
0	0	1	1	0	0	1	0	1	0	1	1	0	1
1	1	0	0	1	1	0	1	1	0	0	1	0	0
0	0	1	0	1	0	1	0	0	1	1	0	1	1
1	1	0	1	0	1	1	0	0	1	0	0	1	0

False: There are only wild tigers in Asia.

203

1	1	0	0	1	0	0	1	1	0	0	1	0	1
0	1	1	0	0	1	1	0	0	1	0	1	1	0
1	0	0	1	1	0	0	1	1	0	1	0	0	1
0	0	1	1	0	1	0	1	1	0	1	1	0	0
1	1	0	0	1	0	1	0	0	1	0	0	1	1
0	0	1	1	0	0	1	1	0	1	0	1	1	0
0	0	1	0	**0**	1	0	1	1	0	1	1	0	1
1	1	0	1	1	0	1	0	0	1	0	0	1	0
0	1	1	0	1	1	0	0	1	0	1	0	1	0
1	0	0	1	0	1	0	1	0	1	0	1	0	1
1	0	1	0	1	0	1	0	1	0	1	0	1	0
0	1	0	1	0	1	1	0	0	1	1	0	0	1
1	0	1	0	1	1	0	1	1	0	0	1	0	0
0	1	0	1	0	0	1	0	0	1	1	0	1	1

True: The classification was first used for roller coasters in 1886.

204

1	0	1	0	1	1	0	1	0	1	0	0	1	0
0	1	0	1	0	1	0	1	0	1	0	1	1	0
0	0	1	1	0	0	1	0	1	0	1	1	**0**	1
1	1	0	0	1	0	1	1	0	1	0	0	1	0
0	0	1	1	0	1	0	1	0	1	1	0	1	0
1	1	0	0	1	1	0	0	1	0	0	1	0	1
0	0	1	0	1	0	1	1	0	1	0	1	0	1
1	1	0	1	0	0	1	0	1	0	1	0	1	0
0	0	1	0	0	1	0	1	1	0	1	0	1	1
0	1	0	1	1	0	1	0	0	1	0	1	0	1
1	0	1	0	1	0	1	0	1	0	1	1	0	0
1	1	0	1	0	1	0	1	0	0	1	0	1	0
0	1	1	0	0	1	0	0	1	1	0	1	0	1
1	0	0	1	1	0	1	0	1	0	1	0	0	1

True

205

1	1	0	1	0	1	0	**0**	1	0	1	0	0	1
1	1	0	1	1	0	0	1	0	0	1	0	1	0
0	0	1	0	1	0	1	1	0	1	0	1	1	0
0	1	1	0	0	1	0	0	1	0	1	1	0	1
1	0	0	1	0	0	1	0	1	0	1	0	1	1
1	1	0	1	1	0	0	1	0	1	0	0	1	0
0	0	1	0	0	1	1	0	1	0	1	1	0	1
1	0	1	0	1	0	0	1	0	1	0	1	1	0
0	1	0	1	0	1	1	0	1	1	0	0	1	0
1	0	1	1	0	0	1	1	0	0	1	0	0	1
0	1	0	0	1	1	0	1	0	1	0	1	0	1
0	0	1	0	0	1	1	0	1	1	0	1	1	0
1	0	0	1	1	0	1	1	0	0	1	0	0	1
0	1	1	0	1	1	0	0	1	1	0	1	0	0

True

206

0	1	1	0	1	0	1	0	0	1	1	0	1	0
1	0	1	0	1	0	1	0	**0**	1	1	0	1	0
0	1	0	1	0	1	0	1	1	0	0	1	0	1
1	0	0	1	0	1	1	0	0	1	0	1	1	0
0	1	1	0	1	0	1	1	0	0	1	0	0	1
0	0	1	1	0	1	0	1	1	0	1	0	1	0
1	1	0	1	1	0	1	0	0	1	0	1	0	0
0	0	1	0	0	1	0	1	1	0	1	0	1	1
0	1	1	0	1	0	0	1	0	1	0	0	1	1
1	0	0	1	1	0	1	0	1	0	1	1	0	0
1	0	0	1	0	1	0	1	1	0	0	1	0	1
0	1	1	0	0	1	1	0	0	1	0	0	1	1
1	1	0	0	1	0	0	1	1	0	1	1	0	0
1	0	0	1	0	1	0	0	1	1	0	1	0	1

True: And maybe the word soldier derives from the Latin sal dare (to give salt).

207

0	1	0	1	0	1	1	0	1	0	0	1	0	1
0	0	1	1	**0**	1	1	0	0	1	1	0	1	0
1	0	1	0	1	0	0	1	0	1	0	1	0	1
1	1	0	0	1	0	1	0	1	0	1	0	1	0
0	1	0	1	0	1	1	0	0	1	0	1	0	1
1	0	1	0	1	1	0	1	0	0	1	0	0	1
1	0	1	1	0	0	1	0	1	1	0	0	1	0
0	1	0	0	1	0	0	1	1	0	1	1	0	1
0	1	0	0	1	1	0	1	0	1	0	1	0	1
1	0	1	1	0	0	1	0	1	0	1	0	1	0
1	1	0	0	1	1	0	1	0	1	0	0	1	0
0	0	1	0	1	0	0	1	1	0	1	1	0	1
1	0	1	1	0	0	1	0	0	1	0	1	1	0
0	1	0	1	0	1	0	1	1	0	1	0	1	0

True

Solutions

Page 9	*A stamp*
Page 75	*Alf=ox, bet=house*
Page 101	*This is a bowline knot.* *The most important and easiest to release knot for making a temporary loop that is reliable.*
Page 135	*H–I–N–O–S–X–Z*
Page 161	*Impossible*
Page 189	*2 to the power of 100 or 1,267,650,600,228,229,401,496,703,205,376 patterns. If you compare the number of brain cells with a grid of 100 billion squares that can have 100 different values then the number of possible patterns is infinite just like your calculator says.*